Alissa M. Carissimi

Interactions among Bacteria and Batrachochytrium dendrobatidis

AF144360

Alissa M. Carissimi

Interactions among Bacteria and Batrachochytrium dendrobatidis

An Investigation of Amphibian Host Immune Response

LAP LAMBERT Academic Publishing

Impressum / Imprint

Bibliografische Information der Deutschen Nationalbibliothek: Die Deutsche Nationalbibliothek verzeichnet diese Publikation in der Deutschen Nationalbibliografie; detaillierte bibliografische Daten sind im Internet über http://dnb.d-nb.de abrufbar.
Alle in diesem Buch genannten Marken und Produktnamen unterliegen warenzeichen-, marken- oder patentrechtlichem Schutz bzw. sind Warenzeichen oder eingetragene Warenzeichen der jeweiligen Inhaber. Die Wiedergabe von Marken, Produktnamen, Gebrauchsnamen, Handelsnamen, Warenbezeichnungen u.s.w. in diesem Werk berechtigt auch ohne besondere Kennzeichnung nicht zu der Annahme, dass solche Namen im Sinne der Warenzeichen- und Markenschutzgesetzgebung als frei zu betrachten wären und daher von jedermann benutzt werden dürften.

Bibliographic information published by the Deutsche Nationalbibliothek: The Deutsche Nationalbibliothek lists this publication in the Deutsche Nationalbibliografie; detailed bibliographic data are available in the Internet at http://dnb.d-nb.de.
Any brand names and product names mentioned in this book are subject to trademark, brand or patent protection and are trademarks or registered trademarks of their respective holders. The use of brand names, product names, common names, trade names, product descriptions etc. even without a particular marking in this work is in no way to be construed to mean that such names may be regarded as unrestricted in respect of trademark and brand protection legislation and could thus be used by anyone.

Coverbild / Cover image: www.ingimage.com

Verlag / Publisher:
LAP LAMBERT Academic Publishing
ist ein Imprint der / is a trademark of
OmniScriptum GmbH & Co. KG
Heinrich-Böcking-Str. 6-8, 66121 Saarbrücken, Deutschland / Germany
Email: info@lap-publishing.com

Herstellung: siehe letzte Seite /
Printed at: see last page
ISBN: 978-3-659-71223-4

Zugl. / Approved by: Memphis, University of Memphis, Master's Thesis, 2015

ACKNOWLEDGEMENTS

Committee members:

Dr. Matthew Parris
Dr. Stephan Shoech
Dr. David Freeman

Vanderbilt University:

Dr. Louise Rollins-Smith
Laura Reinert
Whitney Gammil

James Madison University:

Dr. Reid Harris
Molly Bletz
Eria Rebollar

University of Memphis:

Eleanor Wade
Dr. Jennifer Mandel
Cynthia Stephens
Cassandra Carthon
Marion Lee Barrott
Shane Hanlon
Forrest Brem
Dr. Shin
Steve Kinard
Denita Weeks
Chris Poweless
Dustin Rice
Lyndsey Pierson

ABSTRACT

Carissimi, Alissa M. MS. The University of Memphis. May 2015. Interactions among bacteria and *Batrachochytrium dendrobatidis*: An investigation of amphibian host immune response. Major Professor: Dr. Matthew J. Parris.

The potentially fatal dermatophytic fungus, *Batrachochytrium dendrobatidis* (*Bd*), is one of the most significant contributors to the extensive amphibian declines occurring worldwide. However, some bacteria, such as *Janthinobacterium lividum* and *Pseudomonas fluorescens,* found on amphibian epidermal tissues, water, and soil, effectively inhibit *Bd in vitro*. To determine if these probiotics would promote *Bd* resistance of the southern leopard frog (*Lithobates sphenocephalus*) in the presence of microflora, frogs were inoculated with *J. lividum* and/or *P.fluorescens*. Once colonized by probiotics, *Bd*-induced mortality did not occur when compared to frogs that were not colonized by *J. lividum* and *P. fluorescens*. Those infected with *Bd* prior to bacterial exposure were resistant to colonization and succumbed to chytridiomycosis. Future research must address bioaugmentation strategies using techniques that identify disease associated changes on the epidermis. *J. lividum* and *P. fluorescens* did not colonize *Bd*-infected L. *sphenocephalus*; therefore, are not suitable anti-*Bd* bacteria for this anuran species.

TABLE OF CONTENTS

LIST OF FIGURES

Chapter 1: Exposure of *Lithobates sphenocephalus* to probiotic bacteria for colonization

Introduction

Conceptual Framework

Our understanding of symbioses between animals and microorganisms has developed in recent years, focusing on the application of probiotics to animals for disease mitigation. Amphibians rely partially on epidermal probiotics for defense and immunity against pathogens, such as the chytrid fungus, *Batrachochytrium* dendrobatidis (*Bd*). Their epidermal tissues act as a defense against pathogens, and immune function of this barrier relies significantly on the bacterial community composition. The composition of bacteria on amphibian epidermal tissues includes the probiotic species diversity and relative abundances to one another. The combination of innate immune responses and epidermal bacteria composition significantly influence amphibian disease susceptibility (Bletz et al. 2013; Duda, Vanhoye, & Nicolas, 2002; Jani & Briggs, 2014; Loudon, Woodhams et al. 2014; Rollins-Smith, 1998).

The most significant threat to amphibian populations worldwide is chytridiomycosis, a disease caused by the pathogenic fungus, *Bd* (Crawford, Lips & Bermingham, 2010; Walke & Vredenburg 2008). Infecting its hosts by motile zoospores in the water, *Bd* colonizes and compromises the epidermal tissues of amphibians, causing mortality through cardiac arrest via electrolyte depletion, excessive skin sloughing, and disruption of cutaneous respiration (Voyles et al. 2007). In order to alleviate significant applications of fungicides, and anti-fungal probiotics have been used to rid some amphibians of chytridiomycosis (disease associated with *Bd*) in captivity (e.g., itraconazole; Forzan, Gunn, & Scott, 2008; Harris, Lauer, Simon, Banning, & Alford, 2008; Woodhams *et al.* 2014). Fungicides

1

however, often cause negative side-effects that alter amphibian immunity, growth, and development; therefore, many are not suitable treatments against *Bd* (Brühl, Schmidt, Pieper, & Alscher, 2013; Igbedioh, 1991). The pesticide carbaryl inhibits the release of antimicrobial peptides (AMPs), amphibians' first line of defense against disease, onto their skin (Davidson *et al.* 2007), which can lend them more susceptible to *Bd*. Thus, the use of fungicides in aquatic communities to treat *Bd*, poses a mortality risk to amphibians. Instead of applying detrimental chemicals to treat *Bd*-infected amphibians, non-lethal probiotic bacteria can be used for *Bd* inhibition (Harris et al. 2009; Matz et al. 2012; Muletz, Myers, Domangue, Herrick, & Harris, 2012). Furthermore, the presence of many amphibian pathogens is directly correlated with the properties of probiotic species within their environment, and their relative abundances to other microbes (Jani & Briggs, 2014). If anti-*Bd* probiotics can colonize an amphibian, and are common (i.e., high abundance) in a *Bd* susceptible population, the likelihood of *Bd* remaining in that population is decreased.

Amphibian probiotics isolated from the Red-backed Salamander, *Plethodon cinereus,* among others, protect them from *Bd* infection (Harris et al. 2008). These bacteria, *Janthinobacterium lividum* and *Pseudomonas fluorescens,* produce the metabolites violacein and 2, 4-diacetylphloroglucinol (2, 4-DAPG) that successfully inhibited *Bd* growth on culture plates (Brucker, Baylor et al., 2008; Brucker, Harris et al., 2008). The minimum inhibitory concentration (MIC) of such probiotics is the concentration at which they inhibit 50% of *Bd* growth. Inhibition of *Bd* to concentrations below 10,000 zoospore equivalents (i.e., the mortality threshold for *Bd*) is associated with decreased mortality in amphibians (Vredenburg, Briggs, & Harris, 2011). The MIC for *J. lividum* and *P. fluorescens* has been quantified (Brucker *et al.* 2008; Harris *et al.* 2009; Muletz et al. 2012; Myers, 2012) and therefore, can be used to inoculate hosts at the appropriate concentrations for *Bd* mitigation. However, the MIC for these bacteria toward *Bd* has not been examined in the presence of epidermal microflora on the Southern Leopard Frog, *Lithobates sphenocephalus*. The MIC regarding *Bd* for newly introduced bacteria may be altered in response to the presence of other amphibian epidermal microorganisms and *Bd*, so

2

it is necessary to determine the probiotic MICs for *Bd* in the presence of native (naturally occurring) aquatic and epidermal microorganisms. Equally important, is to test if probiotics need be reapplied if *Bd* susceptible hosts experience repeated exposure to this pathogen.

Literature Review

Bd mitigation strategies should consider the differential immune response of amphibians to exposure. The immune response of *Bd*-infected amphibians varies among species, and one explanation may be the differing metabolites produced by established microorganisms on their epidermis (Woodhams et al. 2007). Probiotic metabolites (violacein, 2, 4-DAPG), directly reduce *Bd* concentrations in inhibitory assays demonstrating that bioaugmentation strategies are feasible (Becker, Brucker, Schwantes, Harris, & Minbiole, 2009; Harris et al. 2009). However, when topically applying *J. lividum* and *P. fluorescens*, the immune response of *L. sphenocephalus* could be stimulated in a manner that could prevent these bacteria from colonizing. The presence of the pathogen *Bd,* however, may facilitate the colonization of introduced probiotic species by altering the amphibians' epidermal community composition (Bletz et al. 2013). For example, *J. lividum* has been shown to increase violacein production on *P. cinereus* epidermal tissues in the presence of *Bd* (Becker et al. 2009) thereby, conferring resistance to *Bd* and increased survivorship compared to unexposed controls. In contrast, other bacteria that are negatively correlated with *Bd* have been shown to decrease as a result of infection; thereby, creating a conundrum (Jani & Briggs, 2014). In order to address this issue, current research must consider both the host and pathogen dynamics for accurate epidemiological measurements.

The mechanisms by which amphibian AMPs and epidermal microflora combat *Bd* are not completely understood. In many amphibian species, the secretion of AMPs from the granular glands is stimulated by both abiotic (e.g., temperature, UV radiation) and biotic (e.g., pathogens, predators) stressors (Duda et al. 2002; Rollins-Smith & Conlon, 2005). The composition of AMP secretions (i.e. peptide families) is species specific and once released onto the epidermal tissues, the AMPs can disrupt the hydrophobic portions of a pathogens' cell membrane, thus killing it (Tennessen et al. 2009). Several families of AMPs, such as brevinin-1 and ranatuerin-2, inhibit the growth of *Bd* thus allowing *Bd* infected hosts to maintain zoospore loads below mortality thresholds (Rollins-Smith & Conlon, 2005). Similarly, Probiotic bacteria, *P. fluorescens* and *J. lividum,* successfully inhibit *Bd* colonization on epidermal tissues once inoculated onto amphibian epidermal tissues (e.g., Red-Legged Frog, *Rana muscosa*) (Harris et al. 2009). *Rana muscosa* does not harbor either species of probiotic bacteria *in vivo,* but both successfully colonized this host; demonstrating the potential for other *Bd* susceptible species to acquire new microbial symbionts. *J. lividum* and *P. fluorescen*s can also act synergistically with endogenously produced AMPs to increase the efficacy of *Bd* inhibition in laboratory settings (Myers et al. 2012). Currently, these have not been confirmed on amphibian epidermal tissues in the presence of resident microflora. The presence of *L. sphenocephalus* microflora may alter *J. lividum* and *P. fluorescens* colonization. Competition, niche availability, and resources are also factors on amphibian tissues that may prevent colonization by newly introduced probiotics (Bletz et al. 2013; Woodhams et al. 2012). Furthermore, the palpable and molecular changes resulting from *Bd* infection will certainly influence microbiome stability.

Bio-augmentation with *P. fluorescens* and *J. lividum* as a strategy to mitigate *Bd* impacts must be considered cautiously. The addition of non-native microbes could create non-target effects in which other species are negatively impacted. However, Scherwinski et al. 2008 found that adding *P. fluorescens* to control the agricultural disease, *Rhizoctonia solani*, was an effective treatment for lettuce with no adverse effects on non-target species. Moriarty (1998) found an increase of prawn survival at an Indonesian farm after introducing *Bacillus harveyi*; this bacterial species decreased the proportion of pathogenic luminous *Vibrio* spp. in the water and sediment. Similarly, bio-augmentation of a North American amphibian species, such as the *L. sphenocephalus,* with *J. lividum* and *P. fluorescens* should pose little risk to non-target organisms given both probiotics are native to North America and commonly

4

found in rivers and soil (Pantanella et al. 2006). Combining both probiotics could serve as a better strategy to control chytridiomycosis *in vivo* because each has a different mechanism of *Bd* inhibition (Duran et al. 2007; Gleeson, O'Gara, & Morrissey, 2010).

Exposing *L. sphenocephalus* to *J. lividum* and *P. fluorescens* in the presence of microflora will establish the most ecologically relevant bio-augmentation strategy; previous research has done so by first removing resident microbes via antibiotics and/or hydrogen peroxide (Becker et al. 2011; Harris et al. 2008). Pre-treatment such as this can cause dysbiosis, in which amphibian microbiota is altered so as to lend them more susceptible to illness and disease (Jani & Briggs, 2014; Woodhams et al. 2014). Vredenburg et al. (2011), topically applied the probiotic *J. lividum* to *Rana muscosa* during an epizootic *Bd* wave without first removing epidermal microbes, and found survivorship of those treated was more than twice that of controls (77% vs. 30%). However, no studies have treated susceptible hosts with multiple bacterial species concurrently both before and after infection by *Bd*. In doing so, conservation efforts and *bd* mitigation strategies can be applied to natural populations, if successful.

L. sphenocephalus is susceptible to *Bd* infection at all life stages, and infections in adults can result in death (Peterson, Wood, Hopkins, Unrine, & Mendonça, 2007; Rothermel et al. 2008).The efficacy of probiotics on *L. sphenocephalus* epidermal tissues can be quantitatively measured by the individuals' response to *Bd* exposure. *L. sphenocephalus* produces brevinin-1 AMPs in response to pathogens (Conlon et al. 2009). Currently, there is no evidence that *J. lividum* or *P. fluorescens* exist on *L. sphenocephalus* epidermal tissues, even though they are sympatric. Bio-augmentation of soil with *J. lividum* and *P. fluorescens* provides protection against *Bd* for other amphibians, even though these bacteria have not been found on their epidermis (Muletz et al. 2012), and by inoculating *L. sphenocephalus* with *J. lividum* and *P. fluorescens* in the presence of resident epidermal microbes, I will test the hypotheses that colonization of *L. sphenocephalus* by these probiotic bacteria is contingent upon microflora and AMP production.

Hypotheses

J. lividum and *P. fluorescens* will not colonize *L. sphenocephalus* epidermal tissues because of inhibition by the combination of AMPs and preexisting microorganisms. The innate immune responses, such as AMPs, will prevent the ready uptake of *J. lividum* and *P. fluorescens* onto *L. sphenocephalus* epidermal tissues. *L. sphenocephalus* individuals collected at Shelby farms, TN tested negative for the presence of these bacterial species; therefore are less likely to acquire new bacterial community members. Furthermore, given *L. sphenocephalus* share habitats with *J. lividum* and *P. fluorescens* found in soil and water, this species would have most likely acquired these bacteria if advantageous to either party and so, it is reasonable to assume the host and bacteria are incompatible. Alternatively, the presence of endogenous AMPs and epidermal microflora may not impact colonization by newly introduced bacteria. These experiments are designed to distinguish between these hypotheses.

Methodology

Animal Husbandry

Four *L. sphenocephalus* egg masses were collected in March 2013 from three ponds and reared in mesocosms at the Meeman Biological Field Station (N35° 21 ', W90°'), Shelby County, Tennessee. Mesocosms simulate a natural pond and ensure individuals are naïve to *Bd* (Rowe & Dunson, 1994). Once individuals underwent metamorphosis (approximately 9-13 weeks), they were transported to the laboratory in containers washed with 10% bleach. Raising *L. sphenocephalus* metamorphs in the lab ensures that individuals are at the same developmental stage and experience the same environmental conditions. Each metamorph was housed individually in 1 L laboratory containers, fed Reptivite©-dusted crickets ad libitum, and water changes

6

were made as needed. Individuals were housed thusly for approximately 20 weeks, at which point frogs were exposed, and capable of producing the full range of adult AMPs (Rollins-Smith, 1998).

Probiotic Cultures

Probiotic suspensions of *J. lividum* and *P. fluorescens* were grown in 1% tryptone broth. Prior to exposing individual frogs to their respective treatments, it was necessary to determine the exponential growth phase for each bacterial species. During the exponential growth phase, bacteria are reproducing at a constant rate, thereby maximizing the chances of successful colonization (Huang, 2013; Monod, 1949). To determine exponential growth for each probiotic, I constructed a standard growth curve using optical density (OD) measurements at one hour intervals for 12 hours after a 24 hour incubation period using a spectrophotometer with wavelength at 590 nm (Sheafor, Davidson, Parr, & Rollins-Smith, 2008). By interpolating each sample to the standard growth curve, I confirmed the concentrations of each probiotic and their exponential growth phase based upon their OD reading. I determined that the concentrations were highest approximately 28 hr after inoculation of 1% tryptone broth media. In addition, I confirmed that after 48 hr, bacterial concentrations steadily declined. The concentrations at 28 hr were much higher than was needed to inoculate each frog therefore; I made serial dilutions and measured the OD of each to confirm which were necessary to achieve 1×10^6 cells/mL. Bacteria were supplied by Dr. Reid Harris at James Madison University, Virginia.

Viable cell counts were conducted by transferring swabs onto tryptone plates with antibiotics, rifampin and penicillin. Both *J. lividum* and *P. fluorescens* were plated on rifampin and penicillin gradients in which the concentration of each antibiotic gradually increased, and resistant colonies from the highest concentration were transferred. Rifampin and penicillin resistant colonies were transferred until the highest antibiotic concentration in the tryptone plates was 1%; a level that exponentially exceeds that found in nature. This ensures isolation of *J. lividum* and *P. fluorescens* from *L. sphenocephalus* epidermal tissues; therefore, these two bacteria

7

can be differentiated from each other based upon morphology. Colony forming units (CFUs, hereafter) of *J. lividum* are characteristically dark purple due to violacein production. *P. fluorescens* is morphologically similar to many species of bacteria; therefore, culturing and selecting for antibiotic resistance of this species was necessary to differentiate it from resident *L. sphenocephalus* microbes.

AMP Collection, Purification, and Efficacy

I isolated AMPs from *L. sphenocephalus* on November 18th, 2013 (one day prior to treatment exposure) by injecting individuals with 10 nM norepinephrine (NE, hereafter) (0.1 mL/g body weight) and placing frogs in a collection buffer (50mM sodium chloride, 25mM sodium acetate, pH 7.0) for 15 min. Once collected, 1% Trifluoroacetic acid was added to each AMP suspension to prevent proteolytic degradation (Rollins Smith et al. 2006) and stored at -4°C. Once thawed, each sample was then filtered through C-18 Sep-Paks (Waters Corporation). Isolated AMPs were then serially diluted and the concentrations from each serial dilution were determined by measuring OD readings and comparing those to the peptide standard, Bradykinin (Micro BCA protein assay kit, Thermo Scientific®). Bovine serum albumin included in the microBCA protein assay kit was not used because amphibian AMPs are more similar in molecular size to the bradykinin peptide. OD readings from *L. sphenocephalus* AMPs are determined by the pigment intensity from the dye; this increases with peptide concentration. OD readings quantified frog AMPs ranging in concentrations from 80 μg/mL to 250 μg/mL. The MIC of *L. sphenocephalus* AMPs against each probiotic was performed via growth inhibition assay; each well in a 96 well micro plate contained a different concentration of AMPs from the serial dilutions of 250, 100, 50, 25, 10, and 5μg/mL. Each dilution was subsequently inoculated under a sterile laminar flow hood with 1 x 10^6 cells/mL of *J. lividum* and *P. fluorescens* independently, and in combination the concentration of each probiotic was 1×10^3 cells/mL. By measuring the optical density of each well at Day 0 and Day 7, I quantified the MIC for each concentration of AMPs against the probiotics. The MIC for AMPs is defined as the concentration by which they inhibit 50 percent of the

growth of microorganisms, such as probiotics and *Bd* zoospores. These MIC values
are analogous to standard LC50 values Eco-toxicologists use to assess the lethal
concentration of a given toxin (i.e., pollutant, insecticide, pesticide etc.) that will
cause 50 % mortality in a wild population (Traunspurger et al. 1997). The MIC of
endogenous *L. sphenocephalus* AMPs against probiotics *J. lividum* and *P. fluorescens*
were quantified. The AMP MIC for *J. lividum* was 200 µg/mL, and for *P.
fluorescens,* it was 150 µg/mL. Therefore, if individuals produce AMP concentrations
that exceed the levels mentioned above, this will most likely prevent *J. lividum* and *P.
fluorescens* from colonizing *L. sphenocephalus.* Although bacteria rapidly reproduce
when transferred to a host or new substrate, if 50 percent of the original inoculum is
inhibited by AMPs, it is less likely that neither probiotic can reach sufficient numbers
to colonize *L. sphenocephalus.*

Probiotic Exposure and Quantification

After quantifying AMP MIC against each bacterial species, AMPs were collected
from resting adults by placing them in the collection buffer after a 2 nM NE injection.
All treatments were prepared in a new sterile container for each frog, and contained a
30mL bacterial suspension with 1% NaCl and sterile tap water. More specifically, at
20 weeks, eight individuals were inoculated with each of the following treatments
during the exponential growth phase for the bacteria: **1)** *J. lividum+/P. fluorescens+*
(i.e., positive control); **2)** *J. lividum+/* heat-killed *P. fluorescens;* **3)** heat-killed *J.
lividum/P. fluorescens+*; and **4)** heat killed *J. lividum* and *P. fluorescens* (i.e.,
negative control). The incubation period for each bacterium was 48 hours to ensure
transmission and/or colonization of each frog. At 48 hours, a fresh pair of nitrile
gloves was used for each individual frog to transfer them into new containers with
fresh water. Each frog was swabbed on days 1, 2, 4, 9, 17, 24, 28, and 35 post
inoculations. Swabs from each frog were placed in 1mL of sterile water and this
water was transferred aseptically to 1% tryptone plates with 10% rifampin. Viable
cell counts were performed 24 hours after each frog was swabbed on days 1, 2, 4, 9,
17, 24, 28, and 35 by counting the number of CFUs on plates that had incubated for
24 hours. The number of CFUs for each probiotic was compared to that from the

original inoculation concentration ($1\text{x} 10^6$ CFU/mL) by converting the number of CFUs into the same unit of measurement for concentration (see Equation).

$$\frac{\text{Number of CFU}}{\text{Volume plated (mL) x total dilution used}} \longrightarrow \frac{\text{Number of CFU}}{\text{mL}} \tag{1}$$

Statistical Analyses

All statistical analyses were performed using SPSS 21 software. One-way Analysis Of Variance (ANOVA) was used to test for differences between and within treatment groups for each day sampled ($N = 40$). Tukey's post-hoc analysis was used to determine which treatment groups significantly differed. All days were then tested together for treatment effects over time using repeated-measures MANOVA for both probiotic samples and AMP production.

Results

The MIC for *L. sphenocephalus* AMPs to inhibit *J. lividum* was 200µg/mL and for *P. fluorescens* was determined to be 150µg/mL. None of the AMPs collected from 2mM NE injections (mild stressor) exceeded this MIC; each individual was inoculated based upon their respective treatment group.

Comparisons of Bacterial Cell Concentrations over Time

MANOVA showed there were no significant differences between treatments with *J. lividum* and the positive control for the duration of the experiment ($F_{1,12} = 1.48, p = 0.403$). The bacterial loads from colonization by *J. lividum* and the positive control did not differ; there were no additive effects resulting from co-culture exposure in treatment group 1 (see Figure 1). Viable cell counts demonstrated the colonization of *L. sphenocephalus* by each bacterial species. Each swab from individuals in bacterial treatment groups yielded a minimum of 100 CFUs. Repeated-measures MANOVA compared the number of probiotic cells for each sample day. Bacterial loads on frogs varied significantly among treatment groups on all sample days ($F_{3,29} = 296.19, p < 0.01$.

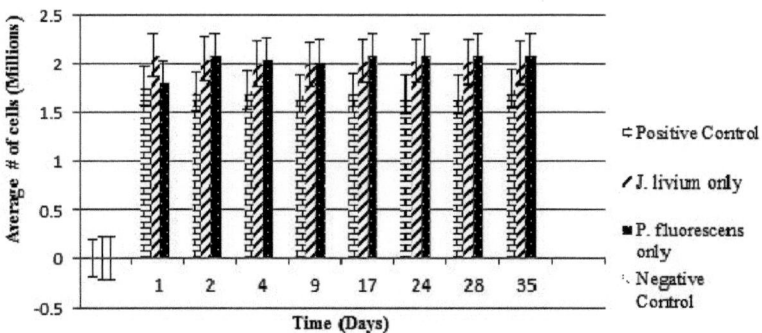

L. sphenocephalus Probiotic Cell Counts

Legend:
- ▭ Positive Control
- ◢ J. livium only
- ▪ P. fluorescens only
- ⸫ Negative Control

X-axis: Time (Days) — 1, 2, 4, 9, 17, 24, 28, 35
Y-axis: Average # of cells (Millions)

Figure 1. Probiotic cell counts on *L. sphenocephalus* epidermal tissues over 35 day period. Treatment x Time was significant at $\alpha < 0.05$.

Multiple comparisons revealed that *L. sphenocephalus* in treatment groups containing *J. lividum* or *P. fluorescens* differed significantly from the negative control (see Figure 1). The concentrations from day 24 onward were similar and time effects were not found when comparing concentrations of each treatment group on day 24 ($p = 0.403$), 28 ($p = 0.374$), and 35 ($p = 0.984$). The fixed concentrations from day 24 onward were most likely the result of probiotic establishment (i.e., symbiosis) with *L. sphenocephalus*. Concentrations on individuals in each of the treatment groups did not decrease below those on day 9; therefore, probiotics colonized because they remained at levels similar to the inoculum concentration at day 0. Tukey's post-hoc analyses showed a significant difference in the number of bacterial cells between those individuals inoculated with either *J. lividum* or *P. fluorescens*. Individuals inoculated with *J. lividum* had higher cell concentrations on average, than those inoculated with *P. fluorescens* ($F_{1,12} = 6.873$, $p = 0.01$). These results suggest that *P. fluorescens* colonized *L. sphenocephalus* at lower cell densities than those of *J. lividum*; abundance differences are observed among many taxa.

AMP production in response to bioaugmentation

Repeated-measures MANOVA confirmed that significant differences in AMP production between treatment groups occurred over time ($F_{1,28} = 11.604$, $p = 0.04$)

(see Figure 2). Introducing *J. lividum* and *P. fluorescens* to a naïve amphibian species mounted a differential immune response, yet AMP production did not inhibit growth or kill these bacteria.

L. sphenocephalus AMP Production

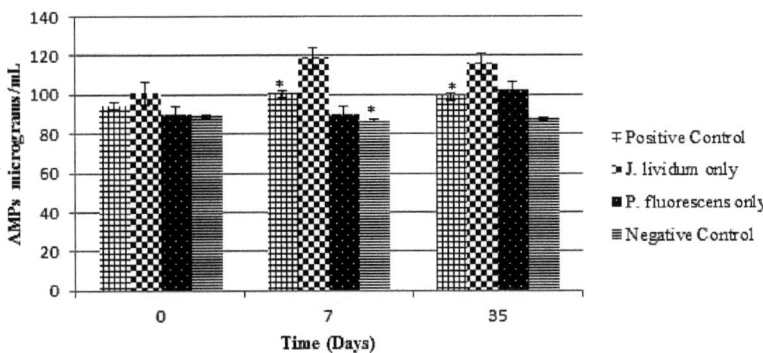

Figure 2. Changes in *L. sphenocephalus* AMP concentrations over 35 day period. Time effects were significant at $F = 11.604$, $P = 0.04$. Significant differences are indicated with an asterisk (*).

Differential AMP production was not observed for individuals in co-culture groups; these frogs did not produce more AMPs than those inoculated with only one bacterial species. There were no significant differences in *L. sphenocephalus* AMP production between any of the treatment groups prior to inoculation with anti-*Bd* bacteria ($F_{3,28} = 0.716, p = 0.557$). Multiple comparisons showed no significant differences for *L. sphenocephalus* AMP production between any bacterial treatment groups on Days 7, and 35 ($p = .352$), this was most likely the result of probiotic establishment. In contrast, the negative control and co-culture groups differed significantly (see Figure 2); roughly 10% AMP production for negative control as compared to co-culture frogs ($p < 0.01$).

Mauchly's test of sphericity was used for within-subjects tests, and no significant differences for AMP production were observed, suggesting that innate immune responses were consistent among individuals ($p > 0.05$). Variations in AMP production have been observed among members of the same species; however, *in vitro* conditions may have eliminated such variation.

Discussion

Findings and Conclusions

The results from this study suggest that *L. sphenocephalus* was successfully colonized by *J. lividum* and *P. fluorescens*; therefore, these probiotics may be effective for preventative and/or treatment regimens against *Bd*. However, the colonization by both bacteria may not be necessary, as there were no additive effects (i.e., increased bacterial loads) on *L. sphenocephalus* epidermal tissues when combining the two. This in part, could be the result of niche availability and niche construction by both bacteria on *L. sphenocephalus* epidermal tissues.

Niche construction is the modification of a particular niche by metabolic processes, physical activities and preferences of the introduced organism; for bacteria such as *J. lividum* and *P. fluorescens,* this would most likely include fixing nutrients (Odling-Smee, Laland, & Feldman, 2003). Fixing nutrients is an important metabolic process common among bacterial symbionts because the host benefits from bacterial colonization by more efficiently obtaining nutrients than otherwise (Postgate, 1998). For example, Rhizobium soil bacteria fix nitrogenous compounds found on the roots of many plants and without these bacteria breaking down these compounds, the plants could not utilize nitrogen in the soil (Sprent, 1987). Amphibian/bacterial symbiosis depends upon several factors, such as nutrient composition, osmolarity, and interactions with previously established resident microbes (Belden & Harris, 2007). Competitive inhibition of microbes occurs quite frequently (Laland & O'Brien, 2010), therefore; it could be that in combination, *J. lividum* and *P. fluorescens* are competing for *L. sphenocephalus* skin resources, thereby preventing additive effects of colonization.

J. lividum may be a better candidate for *Bd* mitigation strategies, as it colonizes in significantly higher concentrations than *P. fluorescens*, which may in turn, reduce the spatial availability for *Bd* zoospores more effectively. The mechanisms by which

13

these probiotics colonized *L. sphenocephalus* are not well understood and further studies should address the probiotic nutrient uptake, possible changes in both the microbial community composition and niche construction on *L. sphenocephalus* epidermal tissues to better understand this complex symbiosis.

The AMP production of *L. sphenocephalus* was shown to increase over time in all treatment groups; this could be due to the growth of the individuals during the experiment. AMP production of an individual is a function of its mass (Rollins-Smith, 2002), and all individuals were less than 2g when inoculated with probiotics and most exceeded this weight by day 35. Conversely, treatment effects were shown to be statistically significant. All but the negative treatment group exhibited increased AMP production until day 7, and then decreased through day 35. One possible explanation is that during the colonization process, immune response was heightened until symbiosis among host and probiotics was established. An overactive immune response to *J. lividum* and *P. fluorescens* may explain the decrease in AMPs over time as a result of an energy trade-off; once AMPs are depleted, it may take up to 60 days to replenish those (Rollins-Smith & Woodhams, 2011). An investigation into the decrease in AMP production should be performed to assess other possible immunological explanations.

P. fluorescens colonized *L. sphenocephalus,* but cell concentrations were not as high as *J. lividum.* It could be possible that *P. fluorescens* can successfully inhibit *Bd* zoospore growth and development to the same extent as *J. lividum,* but at much lower cell concentrations. *P. fluorescens* may serve as a keystone probiotic species, whereby significant contributions to *L. sphenocephalus* immunity can occur in the presence of *Bd* and so, both probiotics should be tested for the prevention and/or treatment of chytridiomycosis. For example, Bletz et al. (2013) found that keystone probiotic species had large effects on amphibian microbial community composition even though cell concentrations were low relative to already established species. *P. fluorescens* could produce larger quantities of anti-*Bd* metabolites than that of other bacterial species. If *P. fluorescens* produces larger quantities than *J. lividum,* the amount of *Bd* inhibition by each bacterium may be similar when *J. lividum* is at significantly higher cell concentrations. The relative abundances of newly introduced probiotics directly effects an amphibians' immune response to pathogens, such as *Bd,* and this is the most likely explanation for the differences observed among and within amphibian populations.

Recommendations for Future Research

Both *J. lividum* and *P. fluorescens* can be topically applied, and if colonization occurs, as it did with *L. sphenocephalus,* probiotics should then be tested in preventative and treatment regimens against *Bd.* More recent findings (Jani & Briggs, 2014) support proposed probiotic colonization as *Bd* alters the microbial communities of infected amphibians, and this change can promote bacterial colonization. More importantly however, is the need to understand the extent to which bacteria can colonize individual amphibians across *Bd* host populations. Future studies can test if probiotics will colonize amphibian hosts despite *Bd*-induced changes in microbial community structure if added to a natural system *in vivo*. Before such experimentation can occur, it is necessary to expose *Bd* susceptible hosts to the disease after anti-*Bd* bacterial colonization has been confirmed in the presence of resident epidermal microbiota.

Chapter 2: Susceptibility to *Bd* after Successful Colonization of
L. sphenocephalus with *J. lividum* and
P. fluorescens

Introduction

Conceptual Framework

Amphibians host a broad range of microorganisms on their epidermis, and most provide fitness advantages via symbioses. Bacteria utilize amphibian resources once colonized onto their epidermal tissues, and in turn, contribute to the amphibians' immunity against pathogens (Jani & Briggs, 2014; Woodhams et al. 2014). Bacteria alone however, may not be enough to prevent, treat, or cure an individual of an exceptionally virulent pathogen. The pathogen *Batrachochytrium dendrobatidis* (*Bd*) is an infectious fungus, responsible for global amphibian declines over the last two decades. *Bd* alters the composition of amphibian epidermal tissues, characteristically hardening (hyperkeratosis) and sloughing off the stratum cornetum and dermis (Berger, Hyatt, Speare, & Longcore, 2005). *Bd* exploits the physical constraints of amphibians (i.e. moist skin for cutaneous respiration) by transmitting infection via motile zoospores in water and moist soil (Johnson & Speare, 2003). This is significant to conservationists who want to implement *Bd* mitigation strategies because amphibians will travel and potentially transport zoospores into new watersheds to hydrate their skin.

Once a *Bd* zoospore reaches a suitable amphibian host, it will attach to the stratum corneum, and develop into a thallus (fruiting body). The thallus then matures into a zoosporangium and expels several new zoospores to either re-infect the current host, or find a new one. At this point the lifecycle for *Bd* is complete and will occur over a 5-7 day period (Berger et al. 2005). *Bd* symptoms can remain undetected in some amphibian species (*Xenopus laevis*) are asymptomatic, and can serve as *Bd* carriers, transmitting infection in susceptible species with which they share habitats (the disease associated with *Bd*) (Rollins-Smith et al. 2005; Voyles et al. 2007). Infected amphibians can succumb to chytridiomycosis (disease associated with *Bd*) within days to months, demonstrating the high inter/intra species variability of symptom severity and mortality.

Literature Review

It is necessary to assess the potential candidates for the prevention of *Bd* transmission and infection because no clear patterns have been observed, thus preventing accurate predictions. However, there is a positive correlation between the survival of *Bd* infected amphibians and the probiotic species present on their epidermal tissues (Harris et al. 2009; Jani & Briggs, 2014; Vredenburg et al. 2011; Woodhams et al. 2014). Many probiotics reside on these tissues because of the permeability and abundant nutrients available to them, similarly, making the amphibian epidermis ideal for infection by pathogens (Bletz et al. 2013; Woodhams et al. 2014). This was demonstrated when amphibians colonized by the probiotic, *Janthinobacterium lividum,* had better survivorship when exposed to *Bd* (Harris et al. 2009). *J. lividum* produces the metabolite violacein which disrupts metabolic function and the structural integrity of *Bd* cells throughout the *Bd* lifecycle (Durán et al. 2007). Furthermore, *J. lividum* produces a biofilm allowing cells to outcompete and kill other microorganisms, such as *Bd,* without harming the host (Pantanella et al. 2007). The probiotic species *Pseudomonas fluorescens* also inhibits *Bd in vitro* by producing the metabolite 2, 4- diacetylphloroglucinol (2, 4-DAPG) (Brucker et al. 2008). *P. fluorescens* is commonly used in agriculture as a successful, low-risk biological control to prevent black rot in U.S. corn crops (Showkat, Murtaza, Laila, & Ali, 2012). Therefore, *J. lividum* and *P. fluorescens* should be considered as preventatives for *Bd* infection in amphibians.

Amphibians have a plethora of microbial species on their epidermal tissues that serve beneficial functions, including, an acquired immune response that contributes to pathogen resistance (Bletz et al. 2013; Jani & Briggs, 2014; Woodhams et al.

17

2014). This collective group of microorganisms living on amphibian tissues makes up the epidermal community composition of that individual. The epidermal community composition of an individual amphibian is highly variable; depending upon brooding behavior, exposure history, and whether a symbiosis can occur. Variance of epidermal community composition can occur within and between amphibian species (McKenzie, Bowers, Fierer, Knight, & Lauber, 2011). For example, isolated populations of the same frog species had different bacterial symbionts depending upon their *Bd* exposure history (Jani & Briggs, 2014). Some amphibian populations are continuously exposed to *Bd,* often during an epizootic wave resulting in massive losses. Such instances make most *Bd* disease parameters (i.e. transmission, susceptibility, infection, and resistance) difficult to predict across and within wild populations. Patterns of *Bd* susceptibility, infectivity, and resistance are not definitive for most amphibian populations because there are vast differences among both host species and *Bd* strains (Berger, Marantelli, Skerratt, & Speare, 2005; Gahl, Longcore, & Houlahan, 2012). However, many probiotics can be used to prevent and treat *Bd* infection in amphibians, if they can colonize, and maintain virulence toward *Bd*. For example, Vredenburg et al. (2011) confirmed that probiotic treatment can be effective if *Bd* infection is low (below 10,000 zoospore equivalents) at the time of treatment.

Bd is an immunosuppressant of susceptible amphibians, and often causes death if infection progresses to 10,000 zoospore equivalents (mortality threshold in susceptible hosts) (Vrendenburg et al. 2011). Symptoms of chytridiomycosis include the restriction of cutaneous gas exchange, skin sloughing, electrolyte, and osmotic imbalance (Brutyn et al. 2012; Voyles et al. 2007). In larval amphibians, despite having a gelatinous body, their keratinized mouthparts become infected with *Bd* resulting in decreased foraging efficiency and developmental stress (Fellers, Reed & Longcore, 2001; Parris & Cornelius, 2004; Venesky, Wassersug, & Parris, 2010). Furthermore, *Bd* zoospores secrete compounds that disrupt cell-to-cell communication in amphibian skin by altering, if not, destroying intercellular junctions (Brutyn et al. 2012). When *Bd* attaches to amphibian epidermal tissues, it initiates the sympathetic nervous system, accompanying norepinephrine (NE) production to stimulate muscle contractions (Rollins-Smith & Conlon, 2005). These muscle contractions increase respiration, heart rate, and oxygen consumption; this signals for antimicrobial peptides (AMPs) to be released from the granular glands in amphibians. AMPs are released onto amphibian epidermal tissues to defend them against invading pathogens (Pask, Cary, & Rollins-Smith, 2013). AMPs produced by amphibians are intraspecific and the arsenal of peptide families varies widely among individuals (Roelants et al. 2013; Tennessen et al. 2009). Many species of ranid frogs have two families of peptides, brevinin-1 and ranatuerin-2; both prevent *Bd* zoospore growth and development when isolated and purified (Rollins-Smith & Conlon, 2005). The mechanisms by which leopard frog AMPs promote disease resistance are well understood (Myers et al. 2012; Pask, Woodhams & Rollins-Smith, 2012; Tennessen *et al.* 2009). As such, the production of the peptide family brevinin-1 is common by the Southern Leopard Frog, *Lithobates sphenocephalus;* and so, this species is the model organism for this study. Amphibian AMP production however does not always result in microorganism death, if it could serve as a symbiont (i.e., probiotic bacteria).

The innate and acquired immune responses of amphibians exhibit a cross-talk that improves response efficacy when threatened by a pathogen, and allow for colonization by probiotics (Bletz et al. 2013; Küng et al. 2014; Richmond, Savage, Zamudio, & Rosenblum, 2009; Woodhams et al. 2007). This cross-talk can occur between *Bd*-inhibitory probiotics (acquired from the environment or topically applied) and endogenously produced immunomodulators, such as AMPs. The combination of AMPs and probiotics is likely the most successful candidate for the prevention of chytridiomycosis in amphibians without using chemically derived anti-fungals. For example, when topically applied to *Plethodon cinereus, J. lividum* concentrations increased as a result of *Bd* exposure (Bletz et al. 2013), suggesting that both the host and symbionts can detect pathogenic threats. Biological controls, such as probiotic bacteria, provide an effective mitigation strategy against fungal diseases of many plants and animals, and can be applied to amphibian systems as well (Jousset, Rochat, Scheu, Bonkowski, & Keel, 2010; Showkat et al. 2012).

Secondary metabolites produced by the bacterial species, *J. lividum* and *P.fluorescens,* effectively inhibit *Bd* growth and reproduction at relatively low concentrations (Becker & Harris, 2010; Bletz et al. 2013; Brucker et al. 2008; Harris et al. 2009; Lauer, Simon, Banning, Lam, & Harris, 2008; Muletz et al. 2012). When *Lithobates sphenocephalus* was exposed to both, *J. lividum* and *P. fluorescens* (Chapter One), AMP production increased, but not enough to inhibit bacterial proliferation; therefore, the probiotics successfully colonized this species. Microorganisms within the epidermal community composition of *L. sphenocephalus* also allowed for colonization by *J.lividum* and *P.fluorescens.* Viable cell counts conducted at several time intervals over a 35 day period revealed that bacterial concentrations remained high enough to inhibit *Bd*. These results suggest that *L. sphenocephalus* is an ideal candidate to test if newly colonized probiotics in conjunction with host AMPs will prevent *Bd*-induced mortality in this species. To further assess the efficacy of *J. lividum* and *P. fluorescens* for *Bd* inhibition, colonized *L. sphenocephalus* will be subsequently exposed to *Bd*.

Hypotheses

J. lividum and *P. fluorescens* will reduce *L. sphenocephalus* susceptibility to *Bd* infection as a result of successful probiotic colonization (Chapter 1). Co-occurring bacterial treatments and endogenously produced AMPs will inhibit *Bd* growth and host mortality more effectively than individual probiotic treatments. This additive effect for *Bd* inhibition was demonstrated when both probiotic metabolites (violacein and 2,4-DAPG), and endogenously produced AMPs were isolated and purified *in vitro* (Myers *et al.* 2012). Furthermore, AMP production of *L. sphenocephalus* will increase as a result of *Bd* exposure, more so than individuals not previously colonized by *J. lividum* and *P. fluorescens.* This experiment will test these hypotheses by experimentally infecting *L. sphenocephalus* previously colonized by *J. lividum* and *P. fluorescens* with *Bd*.

Methodology

Exposure of *L. sphenocephalus* to *Bd*

Eight *L. sphenocephalus* metamorphs (25 weeks old) were used in five treatments for an array of 40 experimental units. All individuals in this experiment were treated and handled in accordance with the approved IACUC protocol #0735. Each metamorph was housed separately, fed twice per week with crickets dusted in Reptivite© vitamins and water was changed as needed. Individuals in treatments 1, 2 and 3 were individuals colonized by probiotics as described in Chapter 1. *Bd* isolate FMB 003 was transferred to 1% tryptone plates and cultured for 7 days to inoculate *L. sphenocephalus.* Several different sub-cultures of this isolate were plated to determine which had the most viable zoospores in each aliquot. Zoospores were stained using malachite green and placed in a hemacytometer. Upon counting zoospores, it was evident that sub-culturing had lensed the pathogen attenuated (Brem, Parris, & Padgett-Flohr, 2013; Langhammer et al. 2013). In a typical 10 µL sample, one would observe hundreds of zoospores; my samples contained, on average, fewer than two dozen. As a result, I reintroduced *Bd* to five *L. sphenocephalus* frogs that were not included in the experiment for two weeks. *Bd* will become virulent after reintroduction onto a new host if it has been sub-cultured for extensive periods of time (Brem et al. 2013). FMB003 did not become more virulent; zoospore concentrations did not increase when re-isolated from *L. sphenocephalus.* Instead, I obtained a fresh frozen stock culture of FMB 003 from Dr. Rollins-Smith at Vanderbilt University in Nashville, TN. This new culture grew exponentially when introduced into 1% tryptone broth, and in 7 days, zoospore counts were increased significantly as compared to the isolate that had been repeatedly sub-cultured. The concentration of zoospores/mL was then calculated at 4 x 10^8 zoospores/mL based upon the number of zoospores from the 10 µL sample within each quadrant of hemacytometer.

Zoospores were then harvested by rinsing *Bd* culture plates with sterile HPLC grade water, and subsequently filtering the rinse into a 10mL syringe. The filtrate was then passed through a 10 μm filter to isolate *Bd* zoospores and to remove thalli/zoosporangia. The optical density was measured in order to determine the concentration of the zoospore suspension. The amount of light that passes through the sample is its optical density (OD, hereafter). The values from the microplate reader are converted into concentrations by comparing them to a standard curve of known concentrations. To construct a standard curve, *Bd* suspensions were serially diluted (full strength, 10^{-1}, 10^{-2}, 10^{-3}, 10^{-4}, 10^{-5}, 10^{-6}, 10^{-7}) and placed in 24 wells (triplicate for each dilution) of a 96-well microplate. OD readings at 590nm of the zoospore solution were used to ensure each frog in the *Bd* treatment group was inoculated with the same *Bd* concentration (1×10^6 zoospores/mL). The *Bd* serial dilution of 10^{-2} was 1×10^6 zoospores/mL; therefore, this dilution of FMB 003 was used for *L. sphenocephalus* individuals in *Bd* treatment groups. Individuals in all treatments, except treatment 4, were inoculated with *Bd* zoospores. Treatments were as follows: 1) co-culture with *J. lividum* and *P. fluorescens*; 2) *J. lividum*; 3) *P. fluorescens* +; 4) Negative control with heat killed *J. lividum, P. fluorescens,* and *Bd*; 5) Positive control with *Bd* only. A fresh pair of nitrile gloves was used to transfer each individual into a 1L container (washed with 10% bleach) with ionized water (1% NaCl).

10mL *Bd* suspensions (1×10^6 zoospores/mL) were transferred, using a pipette, directly into the ionized water containing *L. sphenocephalus* in treatment groups 1, 2, 3, and 5 for 7 days (one *Bd* life cycle) to ensure transmission. After 5 days, a fresh pair of nitrile gloves was used for each individual to remove them from their respective treatments, and placed back into new, clean 1L containers with water.

Probiotic and *Bd* quantification

Every *L. sphenocephalus* in the experiment was swabbed for *Bd* on sample days 2, 4, 9, 17, 24, 35 and 43, until it was no longer detected, or an individual died. Viable cell counts were performed on all sample days, and concentrations for both *J. lividum* and *P.fluorescens* were confirmed until Day 43, at which point all individuals in the positive control group had died. Cell counts were obtained from the probiotics that had been transferred to antibiotic plates from the swabs. Colonies of *J. lividum* and *P. fluorescens* were quantified after a 24 hour incubation period (Miles, Misra, & Irwin, 1938). Swabs were then frozen at -4°C. Once thawed, any DNA present on the swabs was then extracted and purified using Qiagen DNeasy blood and tissue extraction kit©, so that *Bd* could be quantified using qPCR. The purified DNA from each swab was then combined with SYBR Green Master Mix and *Bd* oligonucleotides that were optimum for SYBR Green chemistry. I performed this procedure using methods from previously published research, in accordance with Sigma Aldrich© protocol (Kirshtein, Anderson, Wood, Longcore, & Voytek, 2007; Nolan, Hands, & Bustin, 2006). Six out of eight *L. sphenocephalus* individuals in the *Bd*-positive control group had died before Day 35, consequently; *Bd* and AMP concentrations for only 2 individuals could be determined for the full duration of the experiment. Individuals that died before day 35 were sampled until death, and those that survived were humanely euthanized using MS-222 at the conclusion of the experiment.

AMP Collection, Purification, and Quantification

Each individual in the experiment was rinsed with 50mL of sterile water to remove any transient microbes. To test for changes in AMP concentrations as a result of probiotic colonization and *Bd* exposure, each individual was injected with the appropriate volume of $2\mu g/mL$ of norepinephrine (NE) on days 7 and 35. After NE-injection, each *L. sphenocephalus* was placed in a collection buffer for 15 min to collect their crude peptides. Using 1% Trifluoroacetic acid, each crude peptide mixture was preserved and frozen; this step is necessary to prevent oxidative damage by DNAses, proteases and other enzymes.

After several months, Crude peptide mixtures were thawed to room temperature and after activating sep-paks with buffers A and B, each sample was filtered using column chromatography (Sep-Pak® C18). Each conical containing the purified AMPs were spun down in a centrifuge designed for 50mL falcon tubes, and after 8 hours, was completely dried. In order to quantify the concentrations of each AMP sample, an assay was performed, according to the manufacturer (Micro BCA protein assay kit, Thermo Scientific®). A bradykinin protein standard was serially diluted to concentrations of 400, 200, 50, 25, 10, 5, 2.5, and 1 μg/mL because these are typical concentrations of AMPs collected from *L. sphenocephalus*. The bradykinin peptide was used in place of the bovine serum albumin included in the assay kit because it is structurally similar to AMPs produced by amphibians. AMP samples were then placed into the wells at full strength and half dilution (replicates of three), and averaged using Gen X 5 software. Each 96-well plate was read twice at a wavelength of 570nm to ensure accuracy and minimize any error. Replicate averages with more than 10% variation were excluded and not used for later analyses. All AMP purification and analyses were performed at Vanderbilt University in Nashville, TN.

Quantitative Polymerase Chain reaction (qPCR) Analysis

Frozen *Bd* swabs were analyzed using qPCR and the previously established protocol for SYBR Green Chemistry and *Bd* primers (Longcore et al. 2007; Harris et al. 2009). In order to make such quantifications, *Bd* serial dilutions of 1×10^7, 1×10^6, 1×10^5, 1×10^4, 1×10^3, 1×10^2, 1×10^1, and 1×10^{-1} were used as standards, and were compared to swab samples in the qPCR microplate. The more DNA present, the more binding to SYBR green fluorescent dye, confirming the concentrations of *Bd* zoospores relative to the standard concentrations. Each sample in the qPCR microplate was run in triplicate and those that deviated more than ten percent, were not used for later analysis. After approximately 20 cycles, DNA from the samples was amplified to confirm concentrations of *Bd* zoospores per swab; samples containing minute amounts of *Bd* took as many as 40 cycles for DNA amplification.

Statistical Analyses

Treatment and time effects for *L. sphenocephalus* AMP production were tested with repeated-measures ANOVA using 2 sample days and 5 treatment levels ($N = 34$). Repeated-measures MANOVA was used to determine treatment effects over time for both viable cell counts of probiotics and qPCR zoospore quantification ($N = 36$). Changes in probiotic cell concentrations and *Bd* zoospore loads contained 8 sample days, 2 exposures and 5 treatment levels. One individual from treatments 2 and 3 died, and two individuals in the positive *Bd* control group died as a result of *Bd* exposure by day 5. All individuals that died were preserved in 10% formalin. Harmonic means were used for individuals in *Bd* positive control because only 2 individuals survived beyond one week.

Results

Bd-induced Changes of AMP Production

Using MANOVA, time effects for each treatment group were confirmed, and there were significant differences for *L. sphenocephalus* AMP production on day 7 ($F = 14.141, p < 0.01$) and day 35 ($F = 14.452, p = 0.019$). As seen in figure 3 (next page), AMP production in each of the treatment groups, except for the negative control, differed significantly over time as a result of exposure to *Bd* ($F_{2,3} = 32.121, p < 0.01$).

There were no significant differences in AMP production between individuals with both probiotics/*Bd* and those in the *P. fluorescens*/*Bd* ($F = 14.925, p = 0.219$) or the *J.lividum*/*Bd* groups ($F = 14.925, p = 0.912$). Therefore, AMP production in response to *Bd* exposure was similar regardless of what bacterial species had colonized *L. sphenocephalus* or if *J.lividum* and *P. fluorescens* were combined. There were also significant treatment effects over time the concentrations of AMPs within treatment groups increased in response to *Bd*. Significant differences for *L. sphenocephalus* AMP production were observed on day 7 between groups as a result of treatment exposure ($F_{4,36} = 7.775, p < 0.01$) and between individuals ($F_{8,36} = 31\ p < 0.01$).

Bacterial Concentrations Fluctuates Post-*Bd* Infection

Probiotic cell counts were significantly different among treatment groups for all sample days ($F_{6,20} = 9.390, p = .002$; see Figure 4, next page). Probiotic cell concentrations on *L. sphenocephalus* epidermal tissues differed depending upon the bacterial species and/or combination of species that previously colonized this host.

Time effects occurred for all treatment groups (treatment x time); probiotic concentrations significantly differed between sample days ($F_2 = 5.608$, $p = 0.013$).

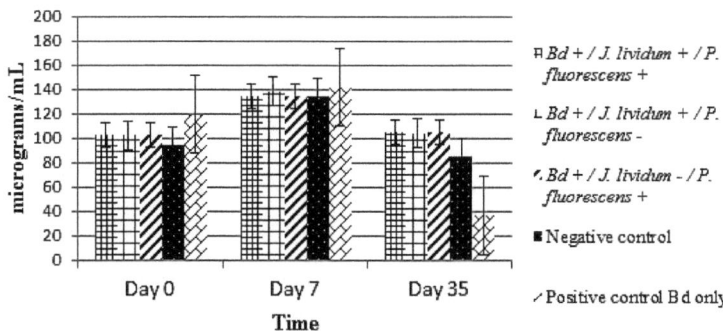

Figure 3. L. sphenocephalus AMP concentrations over 35 day period. Treatment x Time was significant at $\alpha = 0.05$.

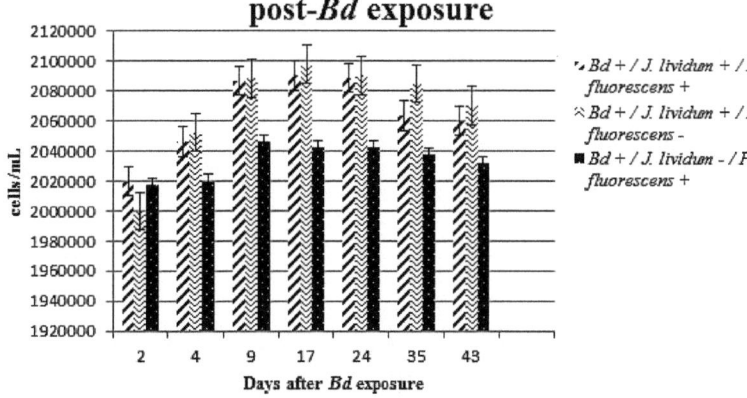

Figure 4. L. sphenocephalus probiotic concentrations over a 43 day period. Treatment x Time was significant on all sample days $P < 0.05$.

In response to *Bd* exposure, probiotic cell concentrations changed over time independent of their presence with one another. There were no significant differences between individuals within probiotic treatment groups containing both probiotics $(F_{1,7} = 2.626, p = 0.123)$ *J. lividum* $(F_{1,7} = 1.758, p = 0.201)$, or *P. fluorescens* $(F_{1,8} = 2.283, p = 0.131)$. The negative control group and the positive control group were excluded from probiotic analyses because they tested negative for both probiotics when swabs were plated on days 2 and 4.

Probiotic Inhibition of *Bd* zoospores

The qPCR results revealed significant differences for *Bd* zoospore loads between treatment groups over time (see Figure 5; $F_{1,4} = 14.494, p < 0.01$). The most significant decrease in *Bd* zoospores on *L. sphenocephalus* was for individuals that had been previously colonized by *J. lividum* (on average reduced by 20 percent by Day 43). Individuals in the positive control maintained *Bd* zoospore concentrations of 1×10^6 zoospores/swab or higher; there were no significant differences between individuals within the positive control group ($F_{1,2} = 3.014, p = 0.095$). Significant differences between individuals previously colonized with *J. lividum* and *P. fluorescens* ($p < 0.01$) were confirmed by Tukey's post-hoc analysis. *L. sphenocephalus* treated with *J. lividum* had significantly less *Bd* zoospores than those treated with *P. fluorescens* ($p = 0.022$). There were no significant differences in *Bd* zoospore loads between those in the *J.lividum* and those in the positive control ($p = 0.772$). All individuals within the negative control group tested negative for the presence of *Bd* zoospores.

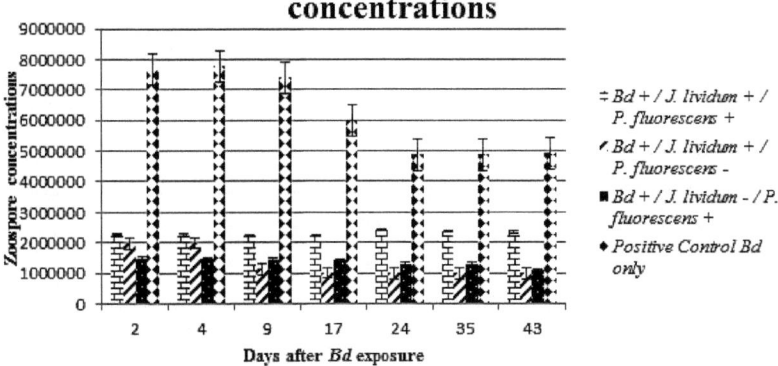

Figure 5. *L. sphenocephalus* zoospore concentrations after colonization by probiotics. All treatment groups were significantly less than the positive control.

Discussion

Discussion of Findings and Conclusions

The results of this study suggest that the use of probiotics for a *Bd* prevention regime can be promising. Both *J. lividum* and *P. fluorescens* lessened the *Bd* mortality rate of *L. sphenocephalus* when compared to the treatment group that was not previously colonized by them. Therefore, *L. sphenocephalus* individuals who have a symbiotic relationship with *J.lividum* and/or *P. fluorescens* are less susceptible to chytridiomycosis. This is also true for wild amphibians that host these bacteria within their epidermal microbial community (Brucker et al. 2008; Culp, Falkinham III, & Belden, 2007; Küng et al. 2014; Lauer, Simon, Banning, Lam, & Harris, 2008). Research findings for amphibian-bacterial symbioses suggest that *J. lividum* and *P. fluorescens* extend the host's immune system by allowing them to acquire some level of resistance to *Bd* (Bletz et al. 2013). For colonization by *J. lividum* and *P.fluorescens* to occur, these bacteria may have competitively inhibited one or more previously established microbes on *L. sphenocephalus* epidermal tissues (Jani & Briggs, 2014; Woodhams et al. 2014). Frogs in this experiment were reared in the laboratory, which may have contributed to the ready uptake of the probiotics after a short duration. It may be that wild *L. sphenocephalus* have a broader, and more

28

complex epidermal community composition containing microbes that could alter colonization by *J.lividum* or *P. fluorescens* (McKenzie et al. 2011).

L. sphenocephalus AMP concentrations did not influence probiotic concentrations, despite the *Bd*-induced increase of AMPs over time. If AMP concentrations exceed the MIC to inhibit *J. lividum* and *P. fluorescens,* the number of bacterial cells for each would have declined significantly or decreased to zero. It may be that the energy requirement for *L. sphenocephalus* AMPs to effectively inhibit probiotics was not necessary for *Bd* inhibition in the presence of epidermal microbes (Garner et al. 2009). Amphibians with relatively low AMP production have substantial amounts of other immune defenses, such as bacterial symbionts, and behavioral adaptations (Harris et al. 2009; Walke, Harris, Reinert, Rollins-Smith, & Woodhams, 2011). Moreover, *L. sphenocephalus* AMP production could be solely the result of *Bd* exposure because *J. lividum* and *P. fluorescens* had colonized this host several weeks prior.

The combination of *L. sphenocephalus* AMPs and probiotics, *J. lividum* and *P. fluorescens,* (treatment group 1) was no more potent against *Bd* than either one of the probiotics (treatment groups 2 and 3). The additive effect in Myers et al. (2012) was observed for purified bacterial metabolites and amphibian AMPs eluted from crude peptides, so other compounds produced by bacterial cells and the crude peptides on *L. sphenocephalus* could have diminished the efficacy of violacein and 2,4-DAPG. The crude peptides of amphibians are comprised of several peptide families, not including AMPs. *L. sphenocephalus* crude peptides could have regulated the bacterial concentrations of probiotics in combination, thereby, preventing them from exceeding some threshold value (Conlon et al. 2009; Pask et al. 2012; Rollins-Smith & Conlon, 2005). It is now understood that the mucosal microbiome of amphibians is often times dictated by the interactions between innate immune responses (AMPs, lymphocytes, and enzymes), and acquired immune responses (exposure history, colonization by microbes, behavior) that may upregulate them (Richmond et al. 2009; Kung et al. 2014; Woodhams et al. 2014).

Another explanation for this experiment as to why there were no additive effects observed when combining *J. lividum* and *P. fluorescens* may be the spatial and nutritional constraints imposed upon these bacteria from *Bd*-induced changes on *L. sphenocephalus* epidermal tissues. The densities of *J. lividum* and *P. fluorescens* may have reached carrying capacity due to limited resources, or competition between the two (Collinge & Ray, 2006). Carrying capacity (defined as *K*), in this context, is the maximum number of bacterial cells that can be sustained by resources on *L. sphenocephalus* epidermal tissues. Such resources can include, niche availability, moisture, and nutrient composition/availability on the epidermal tissues themselves (Jani & Briggs, 2014; Küng *et al.* 2014; Laland & O'Brien, 2010). If the carrying capacity for the probiotics *J. lividum* and *P. fluorescens* has been reached, populations of each naturally decline, potentially rendering them less effective *L. sphenocephalus* immunomodulators for *Bd* infection. Individuals in all treatment groups were experimentally infected with *Bd,* as such; changes of *L. sphenocephalus* epidermal tissues may have restricted bacterial proliferation and additive effects in treatment group 1. *Bd*-infected individuals become depleted of many nutrients, and this would limit the amount available to epidermal microbes (Voyles et al. 2007). The nutrient and spatial availability after *Bd* infection would be similar in all individuals exposed to *Bd*, and this would explain why bacterial densities on *L. sphenocephalus* in treatment group 1 were not higher or more effective than those in groups 2 and 3. Many of the patterns that arise from amphibian symbionts interacting with their immunity can be used to assess future *Bd* mitigation strategies (Woodhams et al. 2014). However, implementation of biological controls (*J. lividum* and *P. fluorescens*) into an amphibian habitat must be done so with caution, and non-target effects must be considered (sympatric organisms may acquire illness from exposure).

The threat of non-target effects can be minimal if *J. lividum* and *P. fluorescens* are to be used in their native geographic locations (i.e. South Eastern United States and Europe). For example, Woodhams et al. (2014) demonstrated the benefit of probiotics *Flavobacterium johnsoniae and P. fluorescens* for treating *Bd*-infected *Alytes obstetricans* (European common midwife toad). Both probiotics pose little risk to other organisms because they are resident on several species of amphibians throughout Europe. Minimizing non-target effects can be achieved if the chosen probiotics have amphibian specificity whereby, other semi-aquatic organisms, such as reptiles and birds, cannot become infected or colonized.

Recommendations for future research

The most suitable *Bd* inhibitors will be those bacteria that reside on amphibian species that have a complex epidermal community composition (Bletz et al. 2013; Woodhams et al. 2014). Many amphibian probiotics serve as immunomodulators; however, they must effectively inhibit *Bd* in the presence of microflora. This research suggests that *J. lividum* and *P. fluorescens* maintain *Bd* inhibition on *L. sphenocephalus* tissues in the presence of resident microbes and AMPs to a similar extent as previous research (Myers et al. 2012). In order to consider *in vivo* bioaugmentation of *L. sphenocephalus,* subsequent studies should address the presence of other organisms within the environment. Anti-*Bd* bacteria must serve a similar function in other organisms they may be able to colonize. It is possible that when an amphibian benefits from a particular bacterial species, a fish could become a new host and that microbe serves as a pathogen, rather than probiotic (Collinge & Ray, 2006). This is often demonstrated in animals with no exposure history to a given microorganism; when they have not yet acquired immunity through crosstalk between the innate and acquired systems.

To facilitate probiotic establishment, whole community level profiling of bacterial populations should be performed on *Bd* susceptible amphibians to ensure that potential *Bd* inhibitory bacteria are compatible with host microflora. Several studies have introduced probiotic bacteria in the presence and absence of *Bd* after the removal of resident microbes (Becker & Harris, 2011; Harris et al. 2009; Vredenburg et al. 2011). "However, this pre-treatment is not always necessary and could remove microbes that facilitate probiotic establishment or contribute to the defense against *Bd"* (Bletz et al. 2013). As such, it is critical to avoid alterations of the microbial community structure that has been established on amphibian epidermal tissues

31

because it may impair immunity. This study indicates that amphibian microflora need not be removed in order to improve immune function. Furthermore, by not removing previously established microbes, this study is the most ecologically relevant approach for inoculating anti-*Bd* bacteria onto amphibian epidermal tissues.

This research can be applied to future studies in which mesocosms can be used to test if *L. sphenocephalus* can be colonized by *J. lividum* and *P. fluorescens* while considering other variables. The presence of other animals, more microorganisms, and different nutrients will influence the outcome when exposing *L. sphenocephalus* to *J. lividum* and *P. fluorescens*. Furthermore, performing this experiment in a mesocosm could address other aspects of amphibian immunity to *Bd* that would not be possible otherwise. *Daphnia magna* (zooplankton) can consume *Bd* zoospores, and is present in waters that *L. sphenocephalus* reside (Hamilton, Richardson, & Anholt, 2012). When present, *D. magna* could alter *L. sphenocephalus* immune responses to *Bd*. Exposure amphibians to organisms found within their natural habitat can be replicated in mesocosm studies, and can provide ecologically relevant information. It is now known that many amphibian microbes cannot be cultured in the lab, but while raising individuals in a mesocosm, we can begin to unravel the complex interactions among host, its symbionts, and *Bd*. Data obtained from this research study can now be used to test if *J. lividum* and *P. fluorescens* are equally effective treatments for *Bd* infection as they are preventatives. In this follow up study, *L. sphenocephalus* would be inoculated *with J. lividum* and *P. fluorescens* after experimental infection by *Bd* is confirmed.

Chapter 3: Addition of *J. lividum* and *P. fluorescens* to

BD-Infected L. sphenocephalus

Introduction

Conceptual Framework

Symbioses are common within a broad range of taxa, and benefit both host organisms and good bacteria (i.e., probiotics). Probiotics can provide advantages to hosts that promote longevity, specifically, by facilitating immune responses (Bletz et al. 2013; Jani & Briggs, 2014). Conservation management plans often utilize probiotics to treat minor infections, diseases, or improve physiological processes, such as digestion in species that are experiencing significant population declines (Hajishengallis, Darveau, & Curtis, 2012; Madsen et al. 2001). "Keystone" probiotics are those that deliver benefits to their host at low concentrations compared to other host microflora. The probiotic keystone hypothesis states that the probiotic need only be added to the organisms' microbial community in low abundance for symbiosis to occur. The probiotic will significantly impact the hosts' microbial community structure and provide some benefit (Hajishengallis, Darveau, & Curtis, 2012). Culturing such probiotics can prevent large losses, if applied to wildlife systems facing an epizootic wave (Bletz et al. 2013). For amphibians, such a bacterium would provide defenses against pathogens by remodeling their epidermal community composition (Jani & Briggs, 2014; Woodhams et al. 2014).

Many amphibian species are susceptible to the fungal disease chytridiomycosis, caused by the highly virulent pathogen, *Batrachochytrium dendrobatidis* (*Bd*). *Bd* is a dermatophytic fungus (requiring keratin for growth) that infects the epidermal tissues (i.e. stratum cornetum, dermis) of its amphibian hosts (Richmond, Savage, Zamudio, & Rosenblum, 2009). Water serves as the primary vector for *Bd* transmission whereby free swimming *Bd* zoospores detect potential hosts via chemotaxis (Rosenblum, Stajich, Maddox, & Eisen, 2008). Once attached to the epidermal tissues of an amphibian host, *Bd* zoospores will mature into zoosporangia. When *Bd* zoosporangia have matured within the epidermis, a discharge papilla forms from which newly formed zoospores are expelled into the water to find keratin (Berger et al. 2005). The threshold for *Bd*-induced mortality is reached once zoospore loads have exceeded 10,000 zoospore equivalents (Vredenburg et al. 2011). Up until this point, susceptible individuals will experience a multitude of disease symptoms associated with chytridiomycosis, such as sloughing off of skin, electrolyte imbalance, and a marked decrease of cutaneous respiration (Voyles et al. 2007). The risk for mortality in *Bd* susceptible amphibians often increases with disease progression because symptoms worsen, and together can cause individuals to undergo cardiac arrest. The epidermal community composition of *Bd*- infected amphibians is also altered with chytridiomycosis progression (Bletz et al. 2013; Jani & Briggs, 2014; Woodhams et al. 2014). Bioaugmentation of amphibian epidermal tissues to prevent/treat infection by *Bd* has been successful using anti-dermatophyte bacteria (Harris et al. 2008; Muletz et al. 2012; Vredenburg et al. 2011; Woodhams et al. 2014).

Amphibians have a microbial assembly on their epidermis and many members produce secondary metabolites and biofilms which inhibit the growth and potential transmission of *Bd* (Bell, 2012; Brucker et al. 2008a; Brucker et al. 2009; Loudon et al. 2014). Individual bacterial cells will uptake beneficial genes (virulence factors) as needed from other neighboring cells. Genes can be readily expelled if unnecessary, to conserve energy. Probiotic virulence factors that protect amphibians from pathogens often depend on the host's exposure history to that particular pathogen. A positive correlation between the concentration of violacein (a toxic metabolite to *Bd*) produced by *Janthinobacterium lividum* and exposure of amphibian epidermal tissues to *Bd* is an example of such a case (Bletz et al. 2013; Muletz et al. 2012). Vertical transmission of such probiotics occurs frequently in amphibians that brood their eggs. While brooding, many *Hyla spp.* and glass frogs transfer bacteria to the outer layer of their gelatinous eggs. As a result, the eggs are less susceptible to fungal, parasitic, and bacterial infections (Walke, Harris, Reinert, Rollins-Smith, & Woodhams, 2011).

Horizontal transmission of probiotics to conspecifics can occur however, most amphibian species host distinct core microbes obtained from the water and soil within their environment (McKenzie et al. 2011; Walke *et al.* 2014).

Literature review

Anti-*Bd* probiotics, *J. lividum* and *Pseudomonas fluorescens,* have successfully colonized amphibian hosts *in vitro* via bioaugmentation (Harris, Lauer, Simon, Banning, & Alford, 2008; Jani & Briggs, 2014; Lauer & Hernandez, 2015; Loudon et al. 2014). Each probiotic produces secondary metabolites that inhibit the growth and development of *Bd* zoospores, resulting in drastic decreases in *Bd* zoospore concentrations on amphibian epidermal tissues. The metabolite violacein produced by *J. lividum* acts on a cellular level to lyse *Bd* zoospores (Pantanella et al. 2007), and 2, 4-Diacetylphloroglucinol (2, 4-DAPG) produced by *P. fluorescens,* effectively slows the development and progression of the *Bd* lifecycle (Harris et al. 2009; Lauer, Simon, Banning, Lam, & Harris, 2008; Muletz et al. 2012). Thus, these bacteria have become a focal point for chytridiomycosis mitigation research. Moreover, *J. lividum* and *P. fluorescens* stimulate the innate immune response of many *Bd* susceptible amphibians in such a way that additive effects can occur when co-cultured. The combination of purified secondary metabolites from co-cultures, and endogenously produced amphibian antimicrobial peptides (AMPs), inhibited *Bd* more effectively (Myers et al. 2012). However, other peptides are secreted from the granular glands in amphibians and so; this study was not an accurate measure of this innate immune response. Amphibian AMPs contribute equally to immune function during *Bd* exposure and must also be considered before bioaugmentation of *Bd*-infected amphibians.

AMP production is the first innate immune response of amphibians, triggered by abiotic and biotic stressors that can facilitate death or disease (Rollins-Smith, 1998). Such abiotic factors include drastic temperature fluctuations, low humidity, and exposure to chemicals. Biotic factors that influence amphibian AMP production include predators, pathogens, and competition. AMP production by amphibians is a

35

"flight or fight" response instigated by the central nervous system. As a result, the adrenal medulla in the amphibian brain stimulates muscle contractions via norepinephrine (NE) production. Contractions of the smooth muscle surrounding amphibian granular glands, stimulates the release of AMPs from an expulsion duct onto the epidermis (Rollins-Smith & Conlon, 2005). Once released, AMPs serve to lyse, degrade, and potentially kill pathogens on the surface of amphibian epidermal tissues, thereby preventing infection, or in some cases, death (Pask et al. 2013; Rollins-Smith & Conlon, 2005, Rollins-Smith, 2009). Due to the immunities AMPs provide an amphibian, colonization of their epidermal tissues by *J. lividum* and *P. fluorescens,* may not be feasible. When selecting anti-dermatophytic bacteria, AMPs produced by the model amphibian species should be considered, and therefore, should not inhibit bacterial growth and colonization.

Bioaugmentation of amphibian epidermal tissues using *J. lividum* and *P. fluorescens* has been a successful *Bd* mitigation strategy for some species (Harris et al. 2009; Becker & Harris, 2011; Loudon, Holland et al. 2014; Vredenburg et al. 2011). This was demonstrated in my current research, of which these probiotics successfully colonized the Southern Leopard Frog, *Lithobates sphenocephalus in vitro.* This is significant because *L. sphenocephalus* is susceptible to *Bd* and prior to this experiment, this species had not been colonized by *J. lividum* and *P. fluorescens* prior to the removal of previously established microbes (See Chapter 1). Pretreatment is not always necessary and could remove microflora that facilitate probiotic establishment or contribute to the defense against *Bd* (Bletz et al. 2013). Following *Bd* exposure, most *L. sphenocephalus* in *J. lividum* and *P. fluorescens* treatment groups did not develop infection. Frogs in these treatments that established *Bd* infection, maintained zoospore loads below the mortality threshold of 10,000 zoospore equivalents (See Chapter 2). This evidence poses a hypothesis, that it is possible that *J. lividum* and *P. fluorescens* could prevent *Bd* transmission. If so, the likelihood of new *Bd* cases would decrease because symptoms during the early stages of *Bd* infection can go undetected. *J. lividum* and *P. fluorescens* target and destroy *Bd* zoospores on *L. sphenocephalus* when applied prior to *Bd* exposure and as such; should be tested as a treatment for chytridiomycosis.

Hypotheses

J. lividum and *P. fluorescens* will successfully colonize *L. sphenocephalus* after infection by *Bd*. In response to *Bd* infection, *L. sphenocephalus* AMP production will be impaired (Küng et al. 2014; Pask et al. 2012; Woodhams et al. 2014), thus allowing for colonization by *J. lividum* and *P. fluorescens*. The presence of *Bd* infection will alter microbial community composition in such a way that newly introduced anti-dermatophyte bacteria, will colonize *L. sphenocephalus* better in co-cultures than separately (Jani & Briggs, 2014; Küng et al. 2014; Lam, Walke, Vredenburg, & Harris, 2010; Woodhams et al. 2014). Equally effective for the treatment and prevention of chytridiomycosis, each bacterium will lessen *Bd* zoospore loads, and prevent mortality after *L. sphenocephalus* is experimentally infected.

Methodology

Bacterial Exposure of *Bd*-Infected L. *sphenocephalus*

Eight individuals in 5 different treatments were housed separately in 1L containers, for a total of 40 ranid frogs. All individuals in this experiment were handled in accordance with the approved IACUC protocol #0735. Frogs were fed live crickets (*Acheta domestica*) twice per week, and sterile tap water (1% NaCl) was changed every 4-7 days. *Bd* isolate FMB003 was transferred to 1% tryptone plates and cultured for 7 days to inoculate *L. sphenocephalus*. On day 7, sterile HPLC grade water was used to rinse *Bd* plates, and the suspension was pushed through a 10mL syringe with a 10 μm filter for zoospore isolation. A separate *Bd* suspension with known concentration (10^6zoospores/mL) was serially diluted (full strength, 10^{-1}, 10^{-2}, 10^{-3}, 10^{-4}, 10^{-5}, 10^{-6}, 10^{-7}) to obtain optical density (OD) readings using a spectrophotometer set at 590nm. These OD values were calculated for each dilution to create a standard curve. The same dilution series for the known sample was used for the FMB003 suspension. The FMB003 suspension was placed into a 96-well microplate to determine which dilution was 10^6 zoospores/mL. This was necessary to confirm that each frog in *Bd* exposure groups was inoculated with the same concentration of zoospores.

By comparing OD values to the standard curve, I confirmed that the FMB003 dilution of 10^{-2} was 1 x 10^6 zoospores/mL. A fresh pair of nitrile gloves was used for each individual to transfer them into a 1L container (washed with 10% bleach) with sterile ionized water (1% NaCl). Using a pipette, 10mL *Bd* suspensions (1 x 10^6 zoospores/mL) were placed directly into the ionized water containing *L. sphenocephalus* in treatment groups 1-4. After 5 days (one *Bd* life cycle), each individual was transferred to their respective treatments into new, clean 1L containers using fresh nitrile gloves.

Once confirmed, *Bd*-positive frogs were then exposed to the following treatments (30mL total volume, 1 x 10^6 bacterial cells/mL) for 48 hours: 1) heat-killed *J. lividum* and *P. fluorescens*; 2) *J. lividum+/P. fluorescens+*; 3) *J. lividum+/P. fluorescens-*; 4) *J. lividum-/P. fluorescens +*. *Bd* negative frogs were placed in treatment 5) negative control with heat-killed *Bd* followed by an application of heat-killed *J. lividum* and *P. fluorescens*. Frogs were placed into new, washed containers with the corresponding treatment for 48 hours to promote bacterial colonization. After 48 hours, all frogs were transferred to separate, clean 1L containers using a fresh pair of nitrile gloves for each individual. All but one individual in the *Bd* positive control group died within one week, and so, there was no data for this group. All individuals that died during the experiment were preserved in 10% formalin. Individuals in the negative control group and those that survived treatment exposure were humanely euthanized after day 35 using MS-222.

Bd and probiotic quantification

Each *L. sphenocephalus* was swabbed on their foot, hand webbing, and ventral surface 10 times using a sterile cotton swab on days 1, 4, 7, 12, 16, 19, 24, and 35.

Swabs were thawed on day 35 for DNA extraction and purification (Qiagen DNeasy blood and tissue extraction kit©), in accordance with the manufacturers protocol. Purified DNA from the swabs was frozen at -4°C until day 35. To quantify *Bd* zoospore concentrations using qPCR, each purified DNA sample was then combined with SYBR Green Master Mix® and *Bd* oligonucleotides (5.8S ribosomal RNA gene) that were optimum for SYBR Green© chemistry. The procedure was performed using methods from previously published research, and in accordance with Sigma Aldrich© protocol (Kirshtein, Anderson, Wood, Longcore, & Voytek, 2007; Nolan, Hands, & Bustin, 2006). If *Bd* was no longer detected on a swab or an individual died, that frog would no longer be used in the analysis.

Immediately after swabbing each frog, swabs were streaked onto plates containing rifampicin and penicillin to obtain viable cell counts for *J lividum* and *P. fluorescens*. Both bacterial species used in this experiment had been sub-cultured to maintain antibiotic resistance (Chapter 1). Colony forming units (CFUs) of *J. lividum* and *P. fluorescens* were counted, and after 24 hours of incubation (26°C), concentrations were quantified (Miles, Misra & Irwin, 1938) for each sample day. Until day 24, viable cell counts from swabs were performed on the same sample days as those for qPCR. On day 24, all individuals in bacterial treatment groups tested negative for *J. lividum* and/or *P. fluorescens*. These frogs were no longer sampled after testing negative for bacteria.

AMP collection, purification and quantification

On days 7 and 35, each frog was rinsed with sterile water and injected with the appropriate volume of 2μg/mL NE to test for changes in AMP concentrations from bacteria and *Bd* exposure. Each individual was then placed in a collection buffer for 15 minutes to collect crude peptides. Using 1% Trifluoroacetic acid (preservative), each crude peptide mixture collected was frozen.

Crude peptide mixtures were purified to isolate AMPs at Vanderbilt University in Nashville, TN. First, the crude peptide mixtures were thawed and after activating Sep-Pak® C18 with methanol, the mixtures were then passed through for purification. Each conical containing the purified AMPs were spun down in a centrifuge designed for 50mL falcon tubes and after 8 hours was completely dried. In order to quantify the concentrations of each AMP sample, an assay (Micro BCA protein assay kit, Thermo Scientific®) was performed using the manufacturer's protocol. A bradykinin protein standard was serially diluted to concentrations of 400, 200, 50, 25, 10, 5, 2.5, and 1 μg/mL were used for the standard because these are typical concentrations of AMPs collected from *L. sphenocephalus*. The bovine serum albumin included in the assay kit was not used because its peptides are significantly larger than any anuran AMPs produced. Each AMP sample was placed into the wells at full strength and half dilution in replicates of three. Plates were incubated at 37°C for two hours to ensure the protein within each AMP sample was detected using the microBCA reagents. Each 96-well plate was read twice at a wavelength of 570nm to ensure accuracy and minimize error. Gen X 5 software was used to average each set of replicates; averages with more than 10% variation were excluded from the analyses. Gen X 5 software data was then transferred to be tested for statistical significance.

Statistical Analyses

All statistical analyses were performed using SPSS 21 software. Treatment and time effects for *L. sphenocephalus* AMP production were tested with repeated-measures ANOVA using 2 sample days and 5 treatment levels ($N = 32$). Repeated-measures MANOVA was used to determine treatment effects over time for both viable cell counts of probiotics and qPCR zoospore quantification ($N = 22$). Changes in probiotic cell concentrations and *Bd* zoospore loads contained 8 sample days, 2 exposures and 5 treatment levels.

Results

Bd infection effects on bacterial colonization

The results for probiotic quantification were not as expected. Despite colonization by probiotics *J. lividum* and *P. fluorescens* on *L. sphenocephalus* epidermal tissues in the preceding experiment, individuals infected with *Bd* prior to bacterial exposure were not colonized by either bacterium. When conducting viable cell counts from swabs using standard protocol (Miles & Misra, 1938), it was determined that *L. sphenocephalus* was highly resistant to colonization by *J.lividum* and *P. fluorescens*. On day 1, only four individuals (treatments 2 and 3) in the experiment had bacterial cell counts similar to the inoculum concentration. By day 4, viable cell counts drastically decreased and eventually reached zero. In treatment group 4, the same trend was observed and by sample day 12, all bacterial treatment groups were negative for viable cell counts. On average, cell counts on day 1 were 15-20 % lower than those from individuals colonized by probiotics prior to *Bd* exposure (Chapter 2). Seven frogs had *J. lividum* and/or *P. fluorescens* on their epidermal tissues after day 1; however, remaining frogs in bacterial treatment groups tested negative for these bacteria (no visible colonies 24 hrs. post-incubation). As a result of small sample sizes for those that did contain *J. lividum* and/or *P. fluorescens* ($N = 7$), multiple comparisons between treatment groups were not made. However, there were significant effects for time (repeated-measures Analysis Of Variance, ANOVA) on bacterial concentrations, as they decreased with each day after application ($F = 386.08, p < 0.01$).

AMP release during *Bd* infection and subsequent bacterial exposure

AMP concentrations increased significantly in response to *Bd* ($F_{3,22} = 9.05$, $p < 0.01$) and the addition of probiotics five days later ($F_{3,22} = 6.609$, $p < 0.01$). No significant differences over time were observed for the negative control ($F = 1.39$, $p = 0.26$). Using repeated-measures MANOVA, it was determined that significant differences occurred between bacterial treatment groups over the duration of the experiment ($F_{3,22} = 517.77$, $p < 0.01$). These results did not include the *Bd* positive control because those individuals had all died; therefore were exempt from analysis. Significant differences for *L. sphenocephalus* AMP production were found between sample days 7 and 35 ($F = 12.35, p = 0.02$).

Bd zoospore loads in response to probiotic intervention

Bd-infected individuals did not have significant decreases in *Bd* zoospore concentrations as a result of bioaugmentation. *L. sphenocephalus* individuals within *J. lividum* and *P. fluorescens* treatment groups had *Bd* zoospore loads similar to one another ($F = 1.32$, $p = 0.41$). All but two of thirty two (6%) individuals within these groups did not die as a result of *Bd* exposure by Day 35. These six frogs that survived *Bd* infection did not contain any *J. lividum* or *P. fluorescens* on their epidermal tissues, suggesting that survival was the result of innate immunity or other resident microbiota. *Bd*-induced changes of the epidermal microbial composition are known to prevent the establishment of new bacteria into the community (Jani & Briggs, 2014). Therefore, the zoospore loads remained at levels to those without probiotics because neither *J. lividum* nor *P. fluorescens* colonized *L. sphenocephalus*. Significant differences in *Bd* zoospore loads were only observed between the negative control group and the bacteria treatment groups on day 1 ($F = 4.57$, $p = 0.04$), and day 4 ($F = 6.52$, $p = 0.03$). All individuals in the *Bd* positive control died within a week as a result of *Bd* zoospore loads exceeding 10,000 zoospore equivalents; thus, zoospore loads in these groups could not be compared to those in probiotic treatment groups for the entire duration of this experiment.

Discussion

Discussion of findings, conclusions

This research has assessed the effects of bioaugmentation of *Bd*-infected *L. sphenocephalus,* and confirmed, *J. lividum* and *P. fluorescens* do not serve as suitable treatments for chytridiomycosis. *L. sphenocephalus* mortality rates declined as a result of probiotic pretreatment (Chapter 2), but in this experiment, *J.lividum* and *P. fluorescens* did not colonize the host, and therefore, did not reduce mortality of *Bd*-infected individuals. There are several possible causes as to why this dichotomy occurred. First, the microbial community structure of *L. sphenocephalus* could have been altered in such a way so as to prevent new bacterial community members on the epidermis (Bletz et al. 2013; Jani & Briggs, 2014; Woodhams et al. 2014). Second, *L. sphenocephalus* epidermal tissues may have become nutritionally or spatially incompatible with the introduced bacteria after *Bd* infection, lensing them ineffective (Becker et al. 2011; Jani & Briggs, 2014; Küng et al. 2014). Third, bacterial colonies could have been removed with sloughed skin cells resulting from *Bd* infection (Jani & Briggs, 2014; Woodhams et al. 2014; Voyles et al. 2007). Based upon the results from this study, it is evident that *J. lividum* and *P. fluorescens* are not keystone probiotic species for *L. sphenocephalus*; they are ineffective *Bd* inhibitors at low concentrations.

Bd-induced increases in AMP production prevented *L. sphenocephalus* colonization by the introduced bacteria; exponential growth of *J. lividum* and *P. fluorescens* was not observed, and their concentrations steadily declined through day 35. Several individuals had AMP concentrations that exceeded 200µg/mL (i.e., above the Minimum inhibitory concentration for both bacteria), and so, this contributed to unsuccessful colonization. The symptoms associated with chytridiomycosis mounted a strong immune response (i.e., high AMP concentrations) for *L. sphenocephalus*; however, other aspects of their innate immunity could have contributed to the death of *J. lividum* and *P. fluorescens*.

Amphibians have many specialized cells and responses that can successfully kill bacteria, especially when the host is faced with a potentially lethal infection (Rollins-Smith, 1998; Savage & Zamudio, 2011; Voyles, Rosenblum, & Berger, 2011). Molecular tests should address the effects of *Bd*-induced immune responses on newly introduced bacteria to understand mechanisms preventing colonization.

Probiotic treatment after *Bd* infection in *L. sphenocephalus* proved less beneficial, likely because *Bd* strain FMB003 was exceptionally virulent (Brem et al. 2013). More virulent strains of *Bd* are characterized by their genetics which encode for faster infection, transmission rates, and increased mortality among susceptible populations (Berger et al. 2005; Richmond et al. 2009). Isolate FMB003 was a fresh stock culture that had only been transferred to culture media three times prior to inoculating *L. sphenocephalus*. Individuals that survived *Bd* exposure, therefore, had high AMP production that prevented probiotic colonization. In this case, *J.lividum* and *P.fluorescens* could not have reproduced at a rate that would have exceeded that of *Bd,* thereby preventing these bacteria from reducing *Bd* zoospore loads below the mortality threshold.

In this research study, 32 individuals were inoculated with *Bd* isolate FMB003 and by day 7, more than half of *L. sphenocephalus* individuals died after exposure. *Bd* strains can have differential effects on both amphibian populations and species (Bataille *et al.* 2013; Berger *et al .*2009). For example, *Lithobates clamitans* was highly susceptible to a Panamanian *Bd* strain, but resistant to the North American strain (Gahl et al. 2012). It is plausible that *L. sphenocephalus* in this experiment were highly susceptible to FMB003 because they could not host *J. lividum* and/or *P. fluorescens* once infected. *L. sphenocephalus* that survived *Bd* infection (18%) were treated with anti-*Bd* bacteria and yet, virtually all of them died, despite this intervention. The susceptibility of *L. sphenocephalus* to *Bd* isolate FMB003 proves that anti-dermatophyte bacteria must be an equally effective treatment as a preventative for chytridiomycosis. Furthermore, amphibian probiotics should be sufficient against more than one *Bd* strain, particularly because of the unpredictable susceptibility both between amphibian populations and among individuals (Gahl et al. 2012). *J. lividum* and *P. fluorescens* are effective inhibitors for some *Bd* strains and amphibian species, but in this experiment, both bacteria were incompetent. *Bd* waves can occur in naïve amphibian populations and infections spread quickly (Lips, Diffendorfer, Mendelson III, & Sears, 2008); therefore, the appropriate probiotics must adequately inhibit *Bd,* regardless of zoospore concentrations found on the amphibian epidermis.

Understanding the epidermal community structure and AMP profile of *Bd* susceptible amphibians is essential for developing feasible bioaugmentation strategies. If infected with *Bd,* an amphibian may suffer a trade-off that could prevent probiotic colonization and/or *Bd* inhibition (Garner et al. 2009). *Bd* infection could result in host dysbiosis whereby the bacterial community no longer contributes to disease resistance or immunity.

Recommendations for future research

Previous research has used growth inhibition assays using purified AMPs and secondary metabolites from anti-*Bd* bacteria to determine MICs against *Bd* using 96-well microplates. This can significantly impact predictions and hypotheses that propose host AMPs and anti-*Bd* bacteria will effectively inhibit *Bd* on the epidermis because other compounds and microorganisms will influence these interactions. By performing inhibition assays using a live host (as opposed to *in vitro*), we can assess accurate measurements of host-pathogen dynamics and probiotic colonization. In such a case, we can determine what host microbes remain effective *Bd* inhibitors after infection (Gibble, Rollins-Smith, & Baer, 2008). Exposure history, heritability, and climate dictate bacterial community composition and AMP profiles of amphibians; therefore, it is necessary to make a complete analysis of their *Bd* inhibition efficacy before choosing an anti-*Bd* probiotic.

New technologies, such as pyrosequencing and whole-community level profiling, have allowed for the identification of bacteria present on amphibians (Bell, 2012; Bletz et al. 2013; Culp et al. 2007; McKenzie et al. 2011). These techniques have demonstrated the subsequent changes of epidermal community composition after *Bd* infection (Jani and Briggs, 2014). However, this data has not been used to test for compatibility between amphibian hosts and newly introduced anti-*bd* bacteria. In doing so, *Bd*-induced discrepancies for bacterial colonization, could be prevented. It may be possible that because these analyses were not performed, the microbial profile on *L. sphenocephalus* epidermal tissues could have changed after *Bd* infection. The relative abundances and proportions of epidermal microbes could have changed in a way so as to induce pathogenicity toward *J. lividum* and/or *P. fluorescens* (Küng et al. 2014; Loudon, Woodhams et al. 2014). It is likely that bacterial densities of those that facilitate *Bd* growth and proliferation increased, while simultaneously inhibiting *Bd* antagonists (Jani & Briggs, 2014). 16-S Pyrosequencing or whole community level profiling could also reveal more anti-*Bd* bacteria that have not been considered previously.

The symptoms associated with chytridiomycosis, such as skin sloughing, electrolyte depletion and osmotic imbalance may have produced an inhospitable environment for *J. lividum* and *P.fluorescens*. The nutritional and spatial requirements for most bacterial species and some strains are highly specified, and even small deviations can be lethal. For example, halotolerant organisms, such as *Staphylococcus spp.,* are found in environments at salt concentrations of 5-25%, and cannot survive outside of this range (Leboffe & Pierce, 2012). Therefore, the *Bd*-induced changes of *L. sphenocephalus* epidermal tissues were most likely outside of the specific ranges for ions, water, and other compounds that *J. lividum* and *P.fluorescens* require. Nonetheless, it could be that epidermal tissues inoculated with *J. lividum* and/or *P. fluorescens* were sloughed off as the disease progressed. *Bd* inhibition by probiotics can depend upon whether the amphibian(s) is infected, and if so, the stage of disease progression at the time of probiotic application.

The amphibian immune response to pathogens may be similar to newly introduced probiotics if the benefit is not direct or rapid. *J. lividum* and *P. fluorescens* are known *Bd* inhibitors, but the imposed immune responses by *L. sphenocephalus* are most likely highly regulated because of infection. Infection of *L. sphenocephalus* by *Bd* may trigger a heightened immune response that once compromised, will not allow for any new members of the microbial community, but rather changes in the abundances of current microbiota. *Bd* mitigation efforts should focus on the alternative mechanisms of immune responses to newly introduced anti-*Bd* bacteria before and after infection. This study revealed a distinctly different immune response to probiotic exposure when *L. sphenocephalus* was infected; supporting the idea that *Bd* influences epidermal community composition (Harris et al. 2009; Rollins-Smith, 2013; Woodhams et al. 2014). If so, how does this change affect the immune response of *Bd*-infected amphibians to other microbes that could promote longevity? This research is the next step toward addressing how the changes imposed upon *Bd*-infected ranid frogs alter the process of bacterial colonization.

REFERENCES

Bangera, M. G., & Thomashow, L.S. (1999). Identification and characterization of a gene cluster fro synthesis of the polyketide antibiotic 2,4-diacytlyphloroglucinol from *Pseudomonas fluorescens* Q2-87. *Journal of Bacteriology* 181, 3155-3163.

Bataille, A., Fong, J. J., Cha, M., Wogan, G. O., Baek, H. J., Lee, H., ... & Waldman, B. (2013). Genetic evidence for a high diversity and wide distribution of endemic strains of the pathogenic chytrid fungus *Batrachochytrium dendrobatidis* in wild Asian amphibians. *Molecular Ecology, 22*(16), 4196-4209.

Becker, M. H., Brucker, R. M., Schwantes, C. R., Harris, R. N., & Minbiole, K. P. (2009). The bacterially produced metabolite violacein is associated with survival of amphibians infected with a lethal fungus. *Applied and Environmental Microbiology, 75(21), 6635-6638.*

Becker, M. H., & Harris, R. N. (2010). Cutaneous bacteria of the redback salamander prevent morbidity associated with a lethal disease. *PLoS One, 5*(6), e10957.

Becker, M. H., Harris, R. N., Minbiole, K. P., Schwantes, C. R., Rollins-Smith, L. A., Reinert, L. K., ... & Gratwicke, B. (2011). Towards a better understanding of the use of probiotics for preventing chytridiomycosis in Panamanian golden frogs. *Ecohealth, 8*(4), 501-506.

Belden, L.K., & Harris, R.N. (2007). Infectious Diseases in Wildlife: The Community Ecology Context. *Frontiers in Ecology and the Environment 5.10,* 533-539.

Bell, S. C. (2012). *The role of cutaneous bacteria in resistance of Australian tropical rainforest frogs to the amphibian chytrid fungus Batrachochytrium dendrobatidis* (Doctoral dissertation, James Cook University).

Berg, G., Grosch, R., & Scherwinski, K. (2007). Risikofolgeabschätzung für den Einsatz mikrobieller Antagonisten: Gibt es Effekte auf Nichtzielorganismen?. *Gesunde Pflanzen, 59*, 107-117.

Berger, L., Hyatt, A. D., Speare, R., & Longcore, J. E. (2005a). Life cycle stages of the amphibian chytrid Batrachochytrium dendrobatidis. *Diseases of Aquatic Organisms, 68*, 51-63.

Berger, L., Marantelli, G., Skerratt, L. F., & Speare, R. (2005b). Virulence of the amphibian chytrid fungus Batrachochytrium dendrobatidis varies with the strain. *Diseases of Aquatic Organisms, 68*, 47-50.

Bletz, M. C., Loudon, A. H., Becker, M. H., Bell, S. C., Woodhams, D. C., Minbiole, K. P., & Harris, R. N. (2013). Mitigating amphibian chytridiomycosis with bioaugmentation: characteristics of effective probiotics and strategies for their selection and use. *Ecology Letters. doi: 10.1111/ele.12099.*

Brannelly, L. A., Chatfield, M. W., & Richards-Zawacki, C. L. (2012). Field and Laboratory Studies of the Susceptibility of the Green Treefrog (*Hyla cinerea*) to *Batrachochytrium dendrobatidis* Infection. *PloS One, 7*, e38473.

Brem, F. M., Parris, M. J., & Padgett-Flohr, G. E. (2013). Re-Isolating *Batrachochytrium dendrobatidis* from an amphibian host increases pathogenicity in a subsequent exposure. *PloS one, 8* (5), e61260.

Briggs, C. J., Vredenburg, V. T., Knapp, R. A., & Rachowicz, L. J. (2005). Investigating the population-level effects of chytridiomycosis: an emerging infectious disease of amphibians. *Ecology, 86,* 3149-3159.

Brucker, R.M., Baylor, C.M., Walters, R.L., Lauer, A., Harris, R.N. & Minbiole, K.P. (2008). The identification of 2,4-diacetylphloroglucinol as an antifungal metabolite produced by cutaneous bacteria of the salamander *Plethodon cinereus. Journal of Chemical Ecology*, 34, 39–43.

Brucker, R. M., Harris, R. N., Schwantes, C. R., Gallaher, T. N., Flaherty, D. C., Lam, B. A., & Minbiole, K. P. (2008). Amphibian chemical defense: antifungal metabolites of the microsymbiont *Janthinobacterium lividum* on the salamander *Plethodon cinereus. Journal of Chemical Ecology, 34* (11), 1422-1429.

Brühl, C. A., Schmidt T., Pieper S., and Alscher A. (2013). Terrestrial pesticide exposure of amphibians: An underestimated cause of global decline? *Scientific reports* 3.

Brutyn, M., D'Herde, K., Dhaenens, M., Van Rooij, P., Verbrugghe, E., Hyatt, A. D & Pasmans, F. (2012). *Batrachochytrium dendrobatidis* zoospore secretions rapidly disturb intercellular junctions in frog skin. *Fungal Genetics and Biology, 49,* 830-837.

Collinge, S. K., & Ray, C. (2006). *Disease ecology: community structure and pathogen dynamics.* Oxford University Press.

Conlon, J. M., Demandt, A., Nielsen, P. F., Leprince, J., Vaudry, H., & Woodhams, D. C. (2009). The alyteserins: Two families of antimicrobial peptides from the skin secretions of the midwife toad *Alytes obstetricans (Alytidae). Peptides, 30,* 1069-1073.

Conlon, J.M., Mechkarska, M., & King, J. D. (2012). Host-defense peptides in skin secretions of African clawed frogs (Xenopodinae, Pipidae). *General and Comparative Endocrinology, 176*, 513-518.

Crawford, A. J., Lips, K. R., & Bermingham, E. (2010). Epidemic disease decimates amphibian abundance, species diversity, and evolutionary history in the highlands of central Panama. *Proceedings of the National Academy of Sciences, 107*(31), 13777-13782

Culp, C. E., Falkinham III, J. O., & Belden, L. K. (2007). Identification of the natural bacterial microflora on the skin of eastern newts, bullfrog tadpoles and redback salamanders. *Herpetologica, 63*, 66-71.

Davidson, C., Benard, M. F., Shaffer, H. B., Parker, J. M., O'Leary, C., Conlon, J. M., & Rollins-Smith, L. A. (2007). Effects of chytrid and carbaryl exposure on survival, growth and skin peptide defenses in foothill yellow-legged frogs. *Environmental Science & Technology, 41*(5), 1771-1776.

Duda, T. F., Vanhoye, D., & Nicolas, P. (2002). Roles of Diversifying Selection and Coordinated Evolution in the Evolution of Amphibian Antimicrobial Peptides. *Molecular Biology and Evolution 19*, 858-864.

Durán, N., Justo, G. Z., Ferreira, C. V., Melo, P. S., Cordi, L., & Martins, D. (2007). Violacein: properties and biological activities. *Biotechnology and Applied Biochemistry, 48*, 127-133.

Fellers, G. M., Green, D. E., & Longcore, J. E. (2001). Oral chytridiomycosis in the mountain yellow-legged frog (*Rana muscosa*). *Copeia, 2001*(4), 945-953.

Forzán, M. J., Gunn, H., & Scott, P. (2008). Chytridiomycosis in an aquarium collection of frogs: diagnosis, treatment, and control. *Journal of Zoo and Wildlife Medicine, 39*, 406-411.

Gahl, M. K., Longcore, J. E., & Houlahan, J. E. (2012). Varying responses of northeastern North American amphibians to the chytrid pathogen Batrachochytrium dendrobatidis. *Conservation Biology, 26*(1), 135-141.

Garcia, T. S., Romansic, J. M., & Blaustein, A. R. (2006). Survival of three species of anuran metamorphs exposed to UV-B radiation and the pathogenic fungus Batrachochytrium dendrobatidis. *Diseases of Aquatic Organisms, 72*, 163-169.

Garner, T. W., Walker, S., Bosch, J., Leech, S., Marcus Rowcliffe, J., Cunningham, A. A., & Fisher, M. C. (2009). Life history tradeoffs influence mortality associated with the amphibian pathogen *Batrachochytrium dendrobatidis*. *Oikos, 118*(5), 783-791.

Gleeson, O., O'Gara, F., & Morrissey, J. P. (2010). The *Pseudomonas fluorescens* secondary metabolite 2, 4 diacetylphloroglucinol impairs mitochondrial function in *Saccharomyces cerevisiae*. *Antonie van Leeuwenhoek*, *97*(3), 261-273.

Gibble, R. E., Rollins-Smith, L., & Baer, K. N. (2008). Development of an assay for testing the antimicrobial activity of skin peptides against the amphibian chytrid fungus *Batrachochytrium dendrobatidis* using *Xenopus laevis*. *Ecotoxicology and Environmental Safety*, *71*, 506-513.

Hajishengallis, G., Darveau, R. P., & Curtis, M. A. (2012). The keystone-pathogen hypothesis. *Nature Reviews Microbiology*, *10*(10), 717-725.

Hamilton, P. T., Richardson, J. M., & Anholt, B. R. (2012). Daphnia in tadpole mesocosms: trophic links and interactions with Batrachochytrium dendrobatidis. *Freshwater Biology*, *57*(4), 676-683.

Harris, R. N., Lauer, A., Simon, M. A., Banning, J. L., & Alford, R. A. (2008). Addition of antifungal skin bacteria to salamanders ameliorates the effects of chytridiomycosis. *Diseases of Aquatic Organisms*, *83*(1), 11.

Harris, R. N., Brucker, R. M., Walke, J. B., Becker, M. H., Schwantes, C. R., Flaherty, D. C., & Minbiole, K. P. (2009). Skin microbes on frogs prevent morbidity and mortality caused by a lethal skin fungus. *The ISME Journal*, *3*, 818-824.

Hinkle, D. E., Wiersma, W. & Jurs, S.G. (2003). Applied statistics for the behavioral sciences. Houghton Mifflin. Boston, Massachusetts.

Huang, L. (2013). Optimization of a new mathematical model for bacterial growth. *Food Control*, *32*, 283-288.

Igbedioh, S. O. (1991). Effects of agricultural pesticides on humans, animals, and higher plants in developing countries. *Archives of Environmental Health: An International Journal*, *46*, 218-224.

Jani, J. A., & Briggs, C. J. (2014). The pathogen Batrachochytrium dendrobatidis disturbs the frog skin microbiome during a natural epidemic and experimental infection. *Proceedings of the National Academy of Sciences*, 201412752.

Johnson, M. L., & Speare, R. (2003). Survival of *Batrachochytrium dendrobatidis* in water: quarantine and disease control implications. *Emerging Infectious Diseases*, *9*, 922-925.

Jousset, A., Scheu, S., & Bonkowski, M. (2008). Secondary metabolite production facilitates establishment of rhizobacteria by reducing both protozoan predation and the competitive effects of indigenous bacteria. *Functional Ecology*, *22* (4), 714-719.

Jousset, A., Rochat, L., Scheu, S., Bonkowski, M., & Keel, C. (2010). Predator-prey chemical warfare determines the expression of biocontrol genes by rhizosphere-associated *Pseudomonas fluorescens*. *Applied and Environmental Microbiology*, *76*, 5263-5268.

Kirshtein, J. D., Anderson, C. W., Wood, J. S., Longcore, J. E., & Voytek, M. A. (2007). Quantitative PCR detection of *Batrachochytrium dendrobatidis* DNA from sediments and water. *Diseases of Aquatic Organisms*, *77*, 11-15.

Köhler, G. A., Assefa, S., & Reid, G. (2012). Probiotic interference of *Lactobacillus rhamnosus* GR-1 and *Lactobacillus reuteri* RC-14 with the opportunistic fungal pathogen *Candida albicans*. *Infectious Diseases in Obstetrics and Gynecology*, *2012*.

Küng, D., Bigler, L., Davis, L. R., Gratwicke, B., Griffith, E., & Woodhams, D. C. (2014). Stability of microbiota facilitated by host immune regulation: informing probiotic strategies to manage amphibian disease. *PloS one*, *9* (1), e87101.

Laland, K. N. & O'Brien, M. J. (2010). Niche Construction Theory and Archaeology. *Journal of Archaeological Method and Theory*, *17.4*, 303-322.

Lam, B. A., Walke, J. B., Vredenburg, V. T., & Harris, R. N. (2010). Proportion of individuals with anti-*Batrachochytrium dendrobatidis* skin bacteria is associated with population persistence in the frog *Rana muscosa*. *Biological Conservation*, *143*(2), 529-531.

Langhammer, P. F., Lips, K. R., Burrowes, P. A., Tunstall, T., Palmer, C. M., & Collins, J. P. (2013). A fungal pathogen of amphibians, *Batrachochytrium dendrobatidis*, attenuates in pathogenicity with in vitro passages. *PloS one*,*8* (10), e77630.

Lauer, A., Simon, M. A., Banning, J. L., André, E., Duncan, K., and Harris, R. N. (2007). Common cutaneous bacteria from the eastern red-backed salamander can inhibit pathogenic fungi. *Copeia* (3), 630–640.

Lauer, A., Simon, M. A., Banning, J. L., Lam, B., and Harris, R. N. (2008). Diversity of cutaneous bacteria with antifungal activity isolated from the female four-toed salamanders. Int. Soc. *Microbial Ecology 2*, 145–157.

Lauer, A., & Hernandez, T. (2014). Cutaneous Bacterial Species from *Lithobates catesbeianus* can Inhibit Pathogenic Dermatophytes. *Mycopathologia*, 1-10.

Leboffe, M. J., & Pierce, B. E. (2012). *Microbiology: Laboratory theory and application*. Morton Publishing Company.

Lips, K. R., Diffendorfer, J., Mendelson III, J. R., & Sears, M. W. (2008). Riding the wave: reconciling the roles of disease and climate change in amphibian declines. *PLoS biology*, *6*(3), e72.

Loudon, A. H., Holland, J. A., Umile, T. P., Burzynski, E. A., Minbiole, K. P., & Harris, R. N. (2014a). Interactions between amphibians' symbiotic bacteria cause the production of emergent anti-fungal metabolites. *Frontiers in Microbiology, 5*.

Loudon, A. H., Woodhams, D. C., Parfrey, L. W., Archer, H., Knight, R., McKenzie, V., & Harris, R. N. (2014b). Microbial community dynamics and effect of environmental microbial reservoirs on red-backed salamanders (Plethodon cinereus). *The ISME journal, 8*(4), 830-840.

Madsen, K., Cornish, A., Soper, P., McKaigney, C., Jijon, H., Yachimec, C., & De Simone, C. (2001). Probiotic bacteria enhance murine and human intestinal epithelial barrier function. *Gastroenterology, 121(3),* 580-591.

Matz, C., Deines, P., Boenigk, J., Arndt, H., Eberl, L., Kjelleberg, S., & Jürgens, K. (2004). Impact of violacein-producing bacteria on survival and feeding of bacterivorous nanoflagellates. *Applied and Environmental Microbiology, 70,* 1593-1599.

McKenzie, V. J., Bowers, R. M., Fierer, N., Knight, R., & Lauber, C. L. (2011). Co-habiting amphibian species harbor unique skin bacterial communities in wild populations. *The ISME Journal, 6* (3), 588-596.

Miles, A. A., Misra, S. S., & Irwin, J. O. (1938). The estimation of the bactericidal power of the blood. *Journal of Hygiene, 38* (06), 732-749.

Monod, J. (1949). The growth of bacterial cultures. *Annual Reviews in Microbiology, 3,* 371-394.

Moriarty, D. J. W. (1998). Control of luminous vibrio species in penaeid aquaculture ponds. *Aquaculture, 164*(1), 351-358.

Muletz, C. R., Myers, J. M., Domangue, R. J., Herrick, J. B., & Harris, R. N. (2012). Soil bioaugmentation with amphibian cutaneous bacteria protects amphibian hosts from infection by *Batrachochytrium dendrobatidis*. *Biological Conservation, 152,* 119-126.

Myers, J. M., Ramsey, J. P., Blackman, A. L., Nichols, A. E., Minbiole, K. P., & Harris, R. N. (2012). Synergistic Inhibition of the Lethal Fungal Pathogen *Batrachochytrium dendrobatidis*: The Combined Effect of Symbiotic Bacterial Metabolites and Antimicrobial Peptides of the Frog *Rana muscosa*. *Journal of Chemical Ecology, 38,* 958-965.

Nolan, T., Hands, R. E., & Bustin, S. A. (2006). Quantification of mRNA using real-time RT-PCR. *Nature Protocols, 1*(3), 1559-1582.

Odling-Smee, F. J., Laland, K. N., & Feldman, M. W. (2003). *Niche construction: the neglected process in evolution* (No. 37). Princeton University Press.

Pantanella, F., Berlutti, F., Passariello, C., Sarli, S., Morea, C., & Schippa, S. (2007) Violacein and biofilm production in *Janthinobacterium lividum*. *Journal of Applied Microbiology, 102*, 992-999.

Parris, M. J., & Cornelius, T. O. (2004). Fungal pathogen causes competitive and developmental stress in larval amphibian communities. *Ecology, 85*, 3385-3395.

Pask, J. D., Woodhams, D. C., & Rollins-Smith, L. A. (2012). The ebb and flow of antimicrobial skin peptides defends northern leopard frogs (*Rana pipiens*) against chytridiomycosis. *Global Change Biology, 18*, 1231-1238

Pask, J. D., Cary, T. L., & Rollins-Smith, L. A. (2013). Skin peptides protect juvenile leopard frogs (Rana pipiens) against chytridiomycosis. *The Journal of Experimental Biology, 216*(15), 2908-2916.

Peterson, J. D., Wood, M. B., Hopkins, W. A., Unrine, J. M., & Mendonça, M. T. (2007). Prevalence of *Batrachochytrium dendrobatidis* in American bullfrog and southern leopard frog larvae from wetlands on the Savannah River site, South Carolina. *Journal of Wildlife Diseases, 43*, 450-460.

Retallick, R. W., & Miera, V. (2007). Strain differences in the amphibian chytrid *Batrachochytrium dendrobatidis* and non-permanent, sub-lethal effects of infection. *Diseases of Aquatic Organisms, 75*(3), 201.

Richmond, J. Q., Savage, A. E., Zamudio, K. R., & Rosenblum, E. B. (2009). Toward immunogenetic studies of amphibian chytridiomycosis: linking innate and acquired immunity. *Bioscience, 59*, 311-320.

Robinson, C. J., Bohannan, B. J., & Young, V. B. (2010). From structure to function: the ecology of host-associated microbial communities. *Microbiology and Molecular Biology Reviews, 74*(3), 453-476.

Roelants, K., Fry, B. G., Ye, L., Stijlemans, B., Brys, L., Kok, P., ... & Bossuyt, F. (2013). Origin and functional diversification of an amphibian defense peptide arsenal. *PLoS genetics, 9*(8), e1003662.

Rollins-Smith, L. A. (1998). Metamorphosis and the amphibian immune system. *Immunological reviews, 166*, 221-230.

Rollins-Smith, L. A., Carey, C., Longcore, J., Doersam, J. K., Boutte, A., Bruzgal, J. E., & Conlon, J. M. (2002). Activity of antimicrobial skin peptides from ranid frogs against *Batrachochytrium dendrobatidis*, the chytrid fungus associated with global amphibian declines. *Developmental & Comparative Immunology, 26*, 471-479.

Rollins-Smith, L. A., & Conlon, J. M. (2005). Antimicrobial peptide defenses against chytridiomycosis, an emerging infectious disease of amphibian populations. *Developmental & Comparative Immunology, 29*, 589-598.

Rollins-Smith, L. A., Woodhams, D. C., Reinert, L. K., Vredenburg, V. T., Briggs, C. J., Nielsen, P. F., & Conlon, M. J. (2006). Antimicrobial peptide defenses of the mountain yellow-legged frog, *Rana muscosa. Developmental & Comparative Immunology, 30*, 831-842.

Rollins-Smith, L. A. (2009). The role of amphibian antimicrobial peptides in protection of amphibians from pathogens linked to global amphibian declines. *Biochimica et Biophysica Acta (BBA)-Biomembranes, 1788*(8), 1593-1599.

Rosenblum, E. B., Stajich, J. E., Maddox, N., & Eisen, M. B. (2008). Global gene expression profiles for life stages of the deadly amphibian pathogen Batrachochytrium dendrobatidis. *Proceedings of the National Academy of Sciences, 105*(44), 17034-17039.

Rothermel, B. B., Walls, S. C., Mitchell, J. C., Dodd Jr, C. K., Irwin, L. K., Green, D. E., & Stevenson, D. J. (2008). Widespread occurrence of the amphibian chytrid fungus *Batrachochytrium dendrobatidis* in the southeastern USA. *Diseases of Aquatic Organisms, 82*(3), 18.

Rowe, C. L., & Dunson, W. A. (1994). The value of simulated pond communities in mesocosms for studies of amphibian ecology and ecotoxicology. *Journal of Herpetology*, 346-356.

Savage, A. E., & Zamudio, K. R. (2011). MHC genotypes associate with resistance to a frog-killing fungus. *Proceedings of the National Academy of Sciences, 108(40),* 16705-16710.

Schulze, A. D., Alabi, A. O., Tattersall-Sheldrake, A. R., & Miller, K. M. (2006). Bacterial diversity in a marine hatchery: balance between pathogenic and potentially probiotic bacterial strains. *Aquaculture, 256*(1), 50-73.

Sheafor, B., Davidson, E. W., Parr, L., & Rollins-Smith, L. (2008). Antimicrobial peptide defenses in the salamander, *Ambystoma tigrinum*, against emerging amphibian pathogens. *Journal of Wildlife Diseases,44*, 226-236.

Showkat, S., Murtaza, I., Laila, O., & Ali, A. (2012). Biological Control of *Fusarium oxysporum* and *Aspergillus* Sp. By *Pseudomonas fluorescens* Isol ated From Wheat Rhizosphere Soil Of Kashmir. *Journal of Pharmacy and Biological Science*s, *1*, 2278-3008.

Sprent, J. I. (1987). *The ecology of the nitrogen cycle.* Cambridge University Press.

Tennessen, J. A., Woodhams, D. C., Chaurand, P., Reinert, L. K., Billheimer, D., Shyr, Y., & Rollins-Smith, L. A. (2009). Variations in the expressed antimicrobial peptide repertoire of northern leopard frog, *Rana pipiens* populations suggest intraspecies differences in resistance to pathogens. *Developmental & Comparative Immunology, 33*, 1247-1257.

Traunspurger, W., Haitzer, M., Höss, S., Beier, S., Ahlf, W., & Steinberg, C. (1997). Ecotoxicological assessment of aquatic sediments with Caenorhabditis elegans (nematoda)—a method for testing liquid medium and whole-sediment samples. *Environmental Toxicology and Chemistry, 16(2),* 245-250.

Van Donk, E., & Hessen, D. O. (1993). Grazing resistance in nutrient-stressed phytoplankton. *Oecologia, 93* (4), 508-511.

Venesky, M. D., Wassersug, R. J., & Parris, M. J. (2010). Fungal pathogen changes the feeding kinematics of larval anurans. *Journal of Parasitology, 96*(3), 552-557.

Voyles, J., Berger, L., Young, S., Speare, R., Webb, R., Warner, J., Rudd, D., Campbell, R. & Skerratt, L. F., (2007). Electrolyte depletion and osmotic imbalance in amphibians with chytridiomycosis. *Diseases of Aquatic Organisms, 77*, 113-118.

Voyles, J., Rosenblum, E. B., & Berger, L. (2011). Interactions between *Batrachochytrium dendrobatidis* and its amphibian hosts: a review of pathogenesis and immunity. *Microbes and Infection, 13*(1), 25-32.

Voyles, J. (2011). Phenotypic profiling of *Batrachochytrium dendrobatidis*, a lethal fungal pathogen of amphibians. *Fungal Ecology, 4*, 196-200.

Vredenburg, V. T., C. J. Briggs, and R. N. Harris (2011). *Host-pathogen dynamics of amphibian chytridiomycosis: the role of the skin microbiome in health and disease.* Washington, DC: The National Academies Press, 2011.

Walke, J. B., Harris, R. N., Reinert, L. K., Rollins-Smith, L. A., & Woodhams, D. C. (2011). Social immunity in amphibians: evidence for vertical transmission of innate defenses. *Biotropica, 43*(4), 396-400.

Walke, J. B., Becker, M. H., Loftus, S. C., House, L. L., Cormier, G., Jensen, R. V., & Belden, L. K. (2014). Amphibian skin may select for rare environmental microbes. *The ISME journal.*

Walker, S. F., Baldi Salas, M., Jenkins, D., Garner, T. W., Cunningham, A. A., Hyatt, A. D., ... & Fisher, M. C. (2007). Environmental detection of *Batrachochytrium dendrobatidis* in a temperate climate. *Diseases of Aquatic Organisms, 77* (2), 105.

Woodhams, D. C., Vredenburg, V. T., Simon, M. A., Billheimer, D., Shakhtour, B., Shyr, Y., & Harris, R. N. (2007). Symbiotic bacteria contribute to innate immune defenses of the threatened mountain yellow-legged frog, *Rana muscosa. Biological Conservation, 138*, 390-398.

Woodhams, D. C., Geiger, C. C., Reinert, L. K., Rollins-Smith, L. A., Lam, B., Harris, R. N., ... & Voyles, J. (2012). Treatment of amphibians infected with chytrid fungus: learning from failed trials with itraconazole, antimicrobial peptides, bacteria, and heat therapy. *Diseases of aquatic organisms, 98*(1), 11.

Woodhams, D. C., Brandt, H., Baumgartner, S., Kielgast, J., Küpfer, E., Tobler, U., & McKenzie, V. (2014). Interacting Symbionts and Immunity in the Amphibian Skin Mucosome Predict Disease Risk and Probiotic Effectiveness. *PloS one, 9* (4), e96375.

THE UNIVERSITY OF
MEMPHIS

IACUC PROTOCOL ACTION FORM

To:	Matthew Parris
From	Institutional Animal Care and Use Committee
Subject	Animal Research Protocol
Date	9-25-13

The institutional Animal Care and Use Committee (IACUC) has taken the following action concerning your Animal Research Protocol No.

Leopard frog Immune Response (0735)

☒ Your proposal is approved for the following period:

From: September 25, 2013 To: September 24, 2016

☐ Your protocol is not approved for the following reasons (see attached memo).

☐ Your protocol is renewed without changes for the following period:

From: To:

☐ Your protocol is renewed with the changes described in your IACUC Animal Research Protocol Revision Memorandum dated for the following period:

From: To:

☐ Your protocol is not renewed and the animals have been properly disposed of as described in your IACUC Animal Research Protocol Revision Memorandum dated

Prof. Guy Mittleman, Chair of the IACUC

Dr. Karyl Buddington, University Veterinarian
And Director of the Animal Care Facilities

STATE OF TENNESSEE
DEPARTMENT OF ENVIRONMENT AND CONSERVATION
Division of Natural Areas
401 Church St., 7ᵗʰ Floor
Nashville, Tennessee 37243-0447
Phone (615) 532-0431, Fax (615)-532-3019

SCIENTIFIC RESEARCH AND COLLECTION PERMIT

Renewal of Permit No.:	Date of Last Annual Review:
Scientific Study Permit No.: **2013-004**	Expires: **December 31, 2013**

Pursuant to the authority granted by Rules 0400-2-2-.23 and 0400-2-8-.30 of the Tennessee Department of Environment and Conservation (TDEC):

Name and Address	Alissa Carissimi	Institution or Organization	University of Memphis

Additional Researcher s	Dr. Matthew Parris	E-mail	crissimi@memphis.edu
		Telephone	**315-745-0126**

The above is permitted to collect **2 to 4** clutches of southern leopard frog eggs (*Lithobates sphenocephalus*) at Meeman-Shelby State Park and State Natural Area.

Details:

Summary of Need: Chytridiomycosis, an emerging infectious disease of amphibians caused by the pathogenic fungus Batrachochytrium dendrobatidis (Bd), has played a major role in amphibian declines over the past two decades. This research examins the effects of probiotic bacterial treatments on the Bd host-pathogen interaction in amphibians

Collection Method(s): Egg masses will be collected by hand or small net and most of the research will occur in the laboratory.

1. Exposure to probiotic bacteria in the absence of disease. Twenty *L. sphenocephalus* metamorphs will be housed under laboratory conditions in each of the four treatment groups for a total of 80 individuals. By rinsing culture plates of each of the bacterial species, Researcher will inoculate tanks in each treatment. Treatments will include: 1- *J. lividum*+/*P. flourescens*+ 2- *J.lividum*+/*P. flourescens*- 3- *J.lividum*-/*P.flourescens*+ 4- Control with heat killed *J.lividum* and *P.flourescens*. Amps will be measured within one hour after one individual comes into contact with inoculated water or soil in the tank. After swabbing the individual to measure relative concentrations of amps and bacteria, it will be euthanized. Two metamorphs will be sampled once per week here after to determine if *J.lividum* and *P. fluorescens* successfully colonized *L. sphenocephalus*.

1a. Susceptibility to *Bd* after successful colonization of *J. lividum* and *P.fluorescens* onto *L. sphenocephalus*. If successfully colonized, researcher will repeat the methods above using the same treatments. In addition, she will inoculate individuals with *Bd* during the exponential growth phases of *J. lividum* and *P. fluorescens* respectively in each treatment. For treatments that contain both *J. lividum* and *P. fluorescens*, researcher will inoculate *L. sphenocephalus* metamorphs during the earlier of two exponential growth phases. During the exponential phase of growth, bacteria are reproducing most rapidly thereby increasing the chances of successful colonization. She will observe changes in concentrations of *J. lividum* and *P. flourescens* both before and after the addition of *Bd* thus, allowing for determination as to whether *Bd* infection persists in the presence of these bacteria.

2. Addition of *J. lividum* and *P. fluorescens* to *Bd* infected *L. sphenocephalus* to determine effects on chytridiomycosis. Researcher will use 20 *L. sphenocephalus* individuals in 5 tanks for a total of 100. Each metamorph will be inoculated with *Bd* and after 5 days (1 *Bd* life cycle) will be exposed to the following treatments: 1- Positive control with *Bd* only. 2- *J. lividum+/P. flourescens+* 3- *J.lividum+/P. flourescens-* 4- *J.lividum-/P.flourescens+* 5- Negative control with heat-killed *Bd*, *J. lividum* and *P. fluorescens*. RESEARCHER will confirm *Bd* infection first by culturing technique. She will swab two individuals from each treatment every 3 days until all individuals are sampled or have died. Concentrations of amps, *Bd*, *J. lividum* and *P. fluorescens* will be recorded and she will test for interactions of *Bd*, bacterial species and effects on amp production. She will also test whether *J. lividum* and *P. fluorescens* can serve as a possible treatment for chytridiomycosis by colonizing an infected host of *Bd*.

Deposition of Specimens: Because the animals will be used for disease-related work and cannot be returned to the wild, they will be humanely euthanized using MS-222.

Subject to the following restrictions: As the researcher wishes to collect eggs of a wildlife species, she must obtain the appropriate permit from the Tennessee Wildlife Resources Agency prior to making any collections. She must also contact the park staff at Meeman-Shelby prior to making any collections. Upon the expiration of this permit, she is to send a progress report and copies of any publication materials to:

Roger McCoy
Division of Natural Areas
401 Church St., 7th Floor
Nashville, TN 37243
Roger.mccoy@tn.gov